Lie Groups and Lie Algebras

A Rewrite of Lie Theory

by

Dennis Morris

Published by: Abane & Right

31-32 Long Row

Port Mulgrave

Saltburn

TS13 5LF

01947 840707

March 2016

Revised May 2016

Contents

Contents

Contents

Contents

Contents

Acknowledgements:

We express our thanks to Howard Georgi for his excellent book:

Lie Algebras in Particle Physics

We express our thanks to K. Moriyasu for his excellent book:

An Elementary Primer for Gauge Theory

Chapter 1

A General Overview

"If we cannot explain something to a twelve-year-old, we do not understand it ourselves." – Richard Feynman (Physics Nobel Laureate)

"I'm not sure I've got Feynman's quotation verbatim, but I've got the gist of it." – Your author (Not Physics Nobel Laureate)

The nature of mathematics is such that everything holds together easily and smoothly. Mathematics and physics is quite simple; only our understanding of it is difficult.

Often, at the frontiers of mathematics and physics, we do not find easy smoothness but instead find complexity and obscurity. We are wont to blame ourselves for the difficulties we have in understanding the subject; we assume our academic inability is the reason we find something difficult. We presume the emperor must be wearing clothes even if we cannot see them. The real reason we find things at the frontiers of knowledge difficult is that there is a lack of clear understanding of these areas within human knowledge. The subject is presented in a difficult and obscure way because the presenter does not clearly understand the subject. This is not to decry the presenter; no-other presenter could do better.

Nor is it always obvious that a particular subject area is at the frontier of human understanding. Lie theory (Lie groups and Lie algebras) is not usually thought of as being at the frontier of human understanding. Lie theory is often seen as being a well matured area of knowledge, and so we ought to be able to understand it without difficulty. Yet for many, including your author when he first met this area of maths, Lie theory is complex and obscure. What is it all about?

Lie groups were invented by the mathematician Sophus Lie (1842 – 1899); they have perplexed students ever since.

Looking at the conventional textbooks on Lie algebra and Lie groups, the student might well feel perplexed by the complications and complexity of it. In most texts, everything seems to be algebraic manipulation with no intuitive understanding of what Lie algebras and Lie groups are really about, where they fit into mathematics, or why we are interested in them. This lack of intuitive understanding makes it difficult for the student to grasp a comprehension of Lie groups in her heart. In short, conventional Lie theory cannot be explained to a twelve-year-old, and therefore, we do not properly understand Lie theory.

If it is true that the complexity and obscurity of conventional Lie theory is due to a failure of our understanding, then where is that failure? Can we rectify this failure?

What is wrong with Lie theory?:
Your author asserts that Lie theory is complicated and obscure because it misunderstands the nature of empty space. Lie theory is about rotations; your author asserts that,

if we misunderstand empty space, then we misunderstand rotation in empty space.

Lie theory assumes all types of empty space are of the same nature as our 4-dimensional space-time. Lie theory completely ignores the very different nature of the spinor spaces even as it speaks of spinors. These spinor spaces are very unlike our 4-dimensional space-time.

Your author has shown elsewhere[1] that our 4-dimensional space-time is a space fabricated from spinor algebras as an emergent expectation space. Such fabricated spaces are very rare, and it seems that our 4-dimensional space-time is unique[2]. The nature of our 4-dimensional space-time is such that we can rotate 2-dimensionally in every 2-dimensional plane (six of them, three 2-dimensional Euclidean planes and three 2-dimensional space-time planes). We call such a space a geometric space meaning we can wave our arms around in it.

Spinor spaces are very different. For example, we can rotate in only three 2-dimensional planes (all Euclidean) within 4-dimensional quaternion space. This effectively constrains rotation in quaternion space to be in one of three intersecting circles. We cannot wave our arms around in quaternion space[3].

The four 3-dimensional spinor spaces allow no 2-dimensional rotations; each of them has only one 3-

---

[1] See: Dennis Morris: Upon General Relativity.

[2] See: Dennis Morris: The Uniqueness of our Space-time.

[3] The perspicacious reader might already be thinking of the intrinsic spin of the electron.

dimensional rotation. Conventional Lie theory ignores the higher dimensional nature of spinor rotations completely.

Lie theory seeks to understand rotations in different types of space. Since conventional Lie theory does not know of the non-geometric nature of rotation in the spinor spaces or of the higher dimensional nature of rotation in the spinor spaces, it is certainly true to say that conventional Lie theory is incomplete. In fact, most of the obscurity and complexity of Lie theory 'melts away' when we introduce an understanding of the spinor spaces and Lie theory becomes explainable to a twelve-year old (a bright one with a month's coaching). Such an explanation requires us to reverse the conventional way of presenting Lie theory.

The conventional presentation of Lie theory begins with a set of matrices called the generators of the Lie algebra, develops that algebra as a set of manipulations, and introduces Lie groups almost as an afterthought. By the time we get to the afterthought, the student is wearied and confused.

In this book, we do things the other way around; we begin with the quite simple intuitive understanding of a Lie group, develop Lie group theory, and then it is the Lie algebra which we derive almost as an afterthought.

Your author opines that this makes Lie groups and Lie algebra much easier to understand. Your author opines that Lie algebra is more properly an afterthought and that the Lie group is the important part of all this 'Lie groups and Lie algebras stuff'. This leads to an unconventional presentation of Lie algebras; it is for the reader to form their own opinion either in favour of or against this way of presentation.

Lie groups are just rotational surfaces:

This book is about continuous groups. Lie group is another name for a continuous group. Roughly, a Lie group is a rotational surface[4].

Technically, a continuous group is a closed set of an uncountable infinite number of associative elements which have inverses and an identity within the set together with an operation which connects any two elements of the set. A continuous group is a differentiable manifold within a space.

For our purposes, we can think of a continuous group as a rotational surface.

Examples of continuous groups are the circle in the 2-dimensional Euclidean plane and the surface of a sphere in 3-dimensional Euclidean space. These are rotation groups in geometric space. A different kind of continuous group is the spinor group of rotations within quaternion space. This is a rotation group in a spinor space[5]. These are the only two types of rotations corresponding to the only two types of space – rotations in geometric space and rotations in spinor space. There are very few geometric space rotations. There are an infinite number of spinor space rotations.

Aside: There are two types of continuous groups; there are continuous rotation groups, rotational surfaces, and there are continuous translation groups. Continuous groups, both

---

[4] There are translation Lie groups, but our main concern is with rotational Lie groups.

[5] A spinor space is a type of complex numbers with one real axis and $(n-1)$ imaginary axes.

types, are called Lie groups. A continuous translation group is just an infinite space, like a 1-dimensional line or a 2-dimensional Euclidean plane. One can translate as much as one likes within such a space without 'falling out' of the space, and, having translated, one can 'un-translate' back to where one began. Of course, one can stay untranslated. The group operation in such a space in translation. In this book, we have no interest in translation groups.

This book is about continuous rotation groups. When we use the phrase 'Lie group' in this book, we will mean a continuous rotation group. The group operation in rotation groups is rotation.

The circle is a continuous rotation group. Every point on the circle is an element of the circle group, known as the Lie group $U(1)$. The Lie group $U(1)$ is isomorphic the Lie group $SO(2)$ which is also the circle group. These group elements, points on the circle, are connected by the act of rotation. The circle group is a rotational surface in both the geometric space, $\mathbb{R}^2$, where it is called $SO(2)$, and in a spinor space, $\mathbb{C}$, where it is called $U(1)$; only rotation groups in the 2-dimensional spaces have this dual nature. Similarly, the surface of a sphere in 3-dimensional $\mathbb{R}^3$ space is a continuous rotation group in which every point is connected by a rotation. This spherical surface is the Lie group known as $SO(3)$. We see that we can think of a continuous rotation group, a Lie group, as a set of points connected by rotation. So, a (rotational) Lie group is a continuous surface within a higher dimensional space which is unit distance from the origin of the space. Without going into detailed proofs, we state that the Lie group is also a differentiable manifold.

Unconventionally, we can forget all about continuous groups and study different types of rotation. We could create a new area of mathematics called, say, 'rotation theory'. We would get the same mathematics as is called (rotational) Lie group theory. It is not unrealistic, (it is unconventional) forgetting the translation Lie groups, to declare that Lie group mathematics is no more than rotation mathematics. Indeed, such a view includes, as an afterthought, Lie algebras as well as Lie groups. Reading the literature, the student will not find this view put so clearly elsewhere. And so, this book is about rotations, or, at least, rotational surfaces in the most general sense.

That most general sense comes down to two, and only two, types of rotational surfaces. One type of rotational surface exists within a geometric space of which our 4-dimensional space-time is the primary example. The other type of rotational surface exists within a spinor space of which quaternion space is an example. These two types of rotational surface are different types of Lie groups with different forms of mathematical expression. Geometric spaces are of the form $\mathbb{R}^n$ in which every axis is a real axis. Spinor spaces have one real axis and $(n-1)$ imaginary axes.

Here's the rub:
The student will now have an intuitive grasp of a Lie group – rotational surface. The interesting aspect of Lie group theory is that there are different types of space in which we

can rotate, and so there are different types of rotation – the difference is not just the dimension of the space. When we study different types of rotation, we are really studying different types of space.

An example of a different type of rotation is rotation in 2-dimensional space-time. Rotation in 2-dimensional space-time is a change of velocity known as a Lorentz boost; this is very different from rotation in the Euclidean plane.

We apologise for being a little repetitive, but once the reader has grasped these concepts, she is well on the way to understanding Lie theory. We repeat; rotation in our 4-dimensional space-time is associated with six 2-dimensional parameters, angles, three Euclidean angles and three space-time angles. In our 4-dimensional space-time, we can rotate in all six 2-dimensional planes. This is why we can wave our arms around in our 4-dimensional space-time. Quaternion space is a 4-dimensional space, but rotation in quaternion space is associated with only three parameters, angles[6]. Thus, in quaternion space, we can rotate in only three of the six 2-dimensional planes. We cannot wave our arms around in quaternion space. In fact, it makes sense to say that there are only three 2-dimensional planes in 4-dimensional quaternion space.

Quaternion space is the space of the intrinsic spin of the electron. The inability to wave our arms around in quaternion space is why electron spin is always either up or

---

[6] This is because the finite group $C_2 \times C_2$ has only three C_2 sub-groups.

down and never in any other direction. Of course, electrons are not the only particles which have intrinsic spin.

The continuous rotation group in quaternion space is the Lie group known as $SU(2)$. $SU(2)$ is something very different from our intuitive view of a spherical surface in $\mathbb{R}^3$ space, but it is still a continuous rotation group[7].

We see that there are as many continuous rotation groups, Lie groups, as there are types of space which allow rotation.

Conventionally, there are more Lie groups than there are types of space because convention accepts rotation in types of space which do not really exist. An example of a non-existent space might be 5-dimensional Riemann space, $\mathbb{R}^5$, with the distance function $dist^2 = a^2 + b^2 + c^2 + d^2 + e^2$. If we had ten angle parameters, this distance function would allow 2-dimensional Euclidean rotation in each of the ten 2-dimensional planes. The only problem is that this space does not exist – proof of non-existence is by mathematics[8] or by simple observation.

Gauge theory implications:
Students of gauge theory will know that forces arise from local variation of the 'rotation phase' in a gauge space. Lie groups, being concerned with the different types of rotation, are directly connected to gauge theory.

---

[7] We will present an intuitive view of rotation in quaternion space in a later chapter.

[8] See : Dennis Morris : The Uniqueness of our Space-time.

Invariance of the distance function:

Rotation is such that it holds invariant the distance function of the space. In 2-dimensional Euclidean space, distance is given by the distance function $dist^2 = x^2 + y^2$. In quantum physics, this distance, which is called the modulus of a complex number, is called probability, and it is an observable quantity of a physical system. Quantum physicists require that this quantity, distance, probability, be unchanged by arbitrary change of co-ordinate system. This number will not be changed by rotation in 2-dimensional Euclidean space.

Sometimes physicists are concerned with the distance function $dist^2 = w^2 + x^2 + y^2 + z^2$. This is the norm of a Weyl spinor which represents an electron with spin[9]. It too is called a probability by physicists. It is also the distance function in quaternion space - $SU(2)$ space, if you prefer.

We see that our continuous rotation groups are intimately connected to the distance function of the space in which we find them and that, at least for the electron, our continuous groups are connected to quantum physics.

The Lie algebra of a Lie group:

A Lie algebra is a set of matrices called generators together with a commutation relation which is a way of combining two of the generators to give a third generator.

---

[9] There are no electrons without spin. However, the Schrödinger equation describes a spinless electron represented by a single complex number.

Conceptually, conventional Lie theory assumes that all spaces are such that they have 2-dimensional rotation in every plane formed from any two of the axial variables; that is, Lie algebra assumes every space is a geometric space like our 4-dimensional space-time. The generators are matrices whose exponential is the 2-dimensional rotation matrix in the appropriate plane; they are so named because they generate the rotation matrix.

Clearly, quaternion space is not of this geometrical space nature. Since Lie algebra assumes that all spaces are geometric, and since not all spaces are geometric, we are going to have to rewrite Lie algebra a little. So we see that we will have to wade through a foggy marsh to understand the quaternion Lie algebra, but that is for later.

Once we have the Lie group, we have the invariance of physics under rotation and we have the locally varying phase of quantum field theory. The Lie algebra does seem like an afterthought.

There is a bit more to it all. How might a being, such as the reader sitting in our 4-dimensional space-time with its 2-dimensional rotations in every 2-dimensional plane, perceive a spinor rotation like the quaternion rotation. We can see only separate 2-dimensional geometric rotations in our 4-dimensional space-time. If we are to see quaternion spinor rotation, then it seems our observation must somehow 'convert' that quaternion spinor rotation into separate 2-dimensional geometric rotations. Perhaps this is where Lie algebra comes in to the picture. In a sense, the Lie algebra generators are the separate 'bits' of the spinor rotation.

This book:

Underlying our approach to Lie groups, there is not only a different view of empty space, but there is also a new approach to mathematics and to physics. We begin this book with an outline of that new approach to mathematics and physics.

Chapter 2

A Different Approach to Mathematics

Circa 2300 BC, Euclid published his elements and thereby formulated mathematics as an axiomatic system.

An axiomatic system is a set of self-consistent axioms which are written in stone and are taken to be inviolably true. Using logical deduction, formally presented as a set of logical axioms, any deductions properly made from the self-consistent axioms are also inviolably true. Absolute truth is normally associated with religion. Mathematical truth is very similar to absolute truth, but mathematics is conditional upon the axioms being true. In practice, the truth of the axioms is never questioned.

The axiomatic system of mathematics contrasts directly with the way science proceeds. Science is based on observation of the universe. Of course, there is an underlying assumption that what scientists observe really exists and is not a figment of their imagination and that the observations scientists make are a true representation of what really exists.

And so we have two diametrically opposed concepts of truth. To mathematics, the axioms are the arbiter of what is true. To physics, and science in general, observation is the arbiter of truth. It is remarkable that physics is written in the language of mathematics in spite of this dichotomy of the nature of truth.

If we ever produce a grand unified field theory of the universe and that theory is written in axiomatic mathematics, then this theory would be fundamentally entwined with the inviolably true axioms which are the invention of mathematicians rather than derived by observation as would be the rest of the theory. Such a fundamental contradiction in the theory would be very unsatisfactory.

Euclid based mathematics on five postulates (axioms). Modern mathematics is based on thirty-four axioms (and four logical axioms). A self-consistent axiomatic system is such that we can add more axioms provided the new axioms are consistent with the older axioms. Sometimes, the new axioms have been included unknowingly from the very start and their inclusion is no more than the statement of a realisation by mathematicians. Other times, entirely new areas of mathematics are opened by adding a new axiom. How do we know this new mathematics is true? It is taken to be true because it is consistent with the established mathematics. How do we know this mathematics really exists? We do not care whether or not this new mathematics really exists in the physical universe; we take it to exist within the axiomatic system we call mathematics.

The thirty-four axioms can be separated into sets such as the metric space axioms, the inner product axioms, the topological axioms, or the algebraic field axioms. Each set is, of course, self-consistent within itself and relates to a particular area of mathematics. There is often mixing of different sets of axioms; for example, the algebraic field axioms are often taken to be included in the metric space axioms by default. In some cases, a subset of axioms forms a set in itself; for example, the four group axioms are four of the seven linear space axioms which are themselves seven of the thirteen division algebra axioms which are themselves thirteen of the fourteen algebraic field axioms.

We see that the axiomatic approach is questionable in two ways.

a) There is an inconsistency between science taking observation to be the arbiter of truth and the mathematics used by science taking the mathematical axioms to be the arbiter of truth. We have two arbiters of truth in the same theory.

b) We do not know that the new mathematics consequent to adding a new axiom really exists. Perhaps the new axiom, and the new mathematics with it, are no more than very sensible figments of the mathematician's imagination.

A new approach to mathematics:

There is another way to do mathematics. In the early part of the twentieth century, Bertrand Russell (1872-1970) proved that the whole of the real numbers can be derived from the existence of the number one. He started with no-more than the number one, and, by logical deduction, showed that the real numbers exist. It took Russell fifteen years to achieve this proof, and the proof runs to many pages (165 in my copy) of logic symbols.

Once we have the real numbers, we can observe that they have properties such as the existence of a multiplicative inverse or the property of multiplicative distributivity over addition. If we list these observed properties, we find we have an exact match for the fourteen algebraic field axioms. Mathematicians will tell you that the axioms of an algebraic field were cast down by thunderbolt from the gods, but, in fact, the axioms were written by mathematicians who were looking over their shoulders at the real numbers. We could take the view that the fourteen algebraic field axioms are no more than observed properties of the real numbers.

We have a new approach to mathematics. We begin with the number one and we declare that mathematics is all that can be deduced from the existence of the number one or observed consequent to such deduction, and mathematics is no more than all that can be deduced or observed from the existence of the number one.

We can take it that the number one exists because if it did not exist there would be zero numbers and zero is a number – it is one number. Alternatively, we can observe the existence of one universe or one electron.

Since the number one exists, then the real numbers exist together with the concepts of multiplication and of addition and of linearity and of associativity and the other observed properties of the real numbers. Further, we can permute different real numbers in many different ways to form the finite groups; for example, the permutations of the three real numbers $\{1, 2, 3\}$ which are $\{123, \ 231, \ 312\}$ form the finite group C_3. We now have all the finite groups from no more than the existence of the number one.

Your author has shown elsewhere[10] that the various types of complex numbers derive from the finite groups and the real numbers. The polar forms of these types of complex numbers each contain a rotation matrix and hence the concept of angle, at least in a purely mathematical sense; the polar forms of these types of complex numbers contain a radial variable and hence the concept of distance.

We will later show that the concept of angle and the concept of distance are insufficient to produce a geometric space[11] in which one can wave one's arms around. We will later show the quaternion space is not a geometric space in which we can wave our arms around. Further, your author has shown elsewhere that the geometric space which is our 4-dimensional space-time is derived from the six 4-dimensional A_3 complex numbers (spinor spaces) and that our 4-dimensional space-time is the unique geometric space. Your author has shown elsewhere that there are no 5-dimensional spaces in which we can wave our arms around, and, indeed, there are no such spaces of any dimension higher than four. This fits with observation, of course.

There is much more that can be derived from no more than the existence of the number one. The list includes, general relativity, classical electromagnetism, the expanding universe, the quantisation of electron spin direction (up or down but never any other direction) and other aspects of physics[12].

---

[10] See the book list at the back of this book.
[11] A geometric space is a space in which we can rotate 2-dimensionally in every 2-dimensional plane.
[12] See the book list at the back of this book.

This new approach to mathematics effectively prunes much modern mathematics from the great tree that has become axiomatic mathematics. Any mathematics which assumes a geometric space of dimension more than four is now pruned from that tree as are any 'geometric' spaces formed from complex axes. There is much more axiomatic mathematics that is likely to be pruned from that great tree by the new approach; possible pruning includes Hilbert spaces and parts of Lie algebra.

It is much easier to deduce new approach mathematics from the number one than it is to prove an area of axiomatic mathematics cannot be deduced from the number one. Previously, your author opined that Clifford algebras do not exist in the new approach mathematics – he was wrong[13]. Previously, your author opined that tensors do not exist in the new approach mathematics – he was wrong in that also.

A new physics:

Physicists see their theories of the universe as being no more than models of the universe constructed by themselves to describe the universe. This is a 'if it works use it' approach, and physicists do not have to worry about the philosophical problems of why the model works. If the model does not work perfectly, physicists can 'tweak' it in some seemingly arbitrary way. Physicists make no claim to understand the universe, they claim only to describe the observed universe.

Your author has shown elsewhere that large amounts of physics (primarily classical physics to date) can be derived

---

[13] See The Naked Spinor by Dennis Morris

from no more than the existence of the number one. (He continues to endeavour to derive the 'missing' bits of physics.) There is a philosophical problem with such derivations. From no more than the number one, your author can calculate a set of squiggly lines on a blackboard which are an exact match for Maxwell's equations[14], but the 'leap' to say "This squiggle is the magnetic field and that squiggle is the electric field" is not in the mathematics. Why is it that something with no more tangibility than a number or a finite group appears to us to be as solid as your author's head? The modelling physicist does not have this problem because she observes the electric and magnetic fields and uses a squiggle to represent them. The modelling physicist knows that her squiggles are not really the electric and the magnetic fields. The 'derive everything from number one' physicist asserts that mathematics is the universe, but she cannot explain the 'leap' to solidity. Having said that, looking at particle physics, it does seem than everything is just empty space. The fundamental particles seem to have no size or solidness of their own. Of course, we get the empty spinor spaces from the number one and the empty geometric space from the spinor spaces, and so we hope that the 'leap' to solidity will one day be understood.

Thus we have a new way of doing physics. We derive physics from the existence of the number one. This is, of course, exactly the new approach to mathematics, and so physics and mathematics are the same thing. We have ridded ourselves of the dichotomy of two different arbiters of truth in the same theory.

---

[14] See The Physics of Empty Space by Dennis Morris

There are many other advantages to this new approach to physics. Since the universe is no more than a manifestation of the number one, and since the number one has to exist, we now know why the universe exists. We can answer other questions like why we have the observed number of physical constants in the universe or why we have two classical forces, that is gravity and electromagnetism, rather than, say, three classical forces or perhaps only one classical force. We know why our space-time is 4-dimensional and why it has the distance function it does have. We have explanations, which might be correct, for dark energy, three generations of particles, and the breaking of parity by the weak force; it is a little early to claim the correctness of these explanations.

The new approach physics is far from mature. There is still much that is not understood and much physics which has not yet been derived. We therefore need to keep the traditional physics for now; we cannot throw away something that works so well until we have something at least as good with which to replace it. However, the new approach physics does seem to be better than the traditional physics in the area of understanding empty space and continuous groups. We adopt the new approach physics in this book because this book is concerned with continuous groups – that is Lie groups to give them their Sunday name.

Lie groups from the number one:

Continuous groups are called Lie groups. Our approach to these Lie groups is from a very different perspective to the approach, via Lie algebra, normally presented to students. We will derive all the continuous groups from no more than the number one (via the real numbers and the finite groups). Having obtained the continuous groups, Lie groups, we will derive the Lie algebra associated with that group. There are many surprises in store, for although we obtain all but one of the standard Lie algebras used in physics, we are missing one. We obtain $U(1)$, and $SU(2)$, and $SO(3)$ and the Lorentz group $SO(1,3) \cong SO(3,1)$, but we cannot find $SU(3)$ anywhere.

Note: *For this approach to Lie groups, your author is much indebted to Dr Frederic P. Schuller for his lectures on the geometric anatomy of theoretical physics as presented free of charge through iTunes. If the reader has not seen lectures on iTunes, your author recommends them to him. In particular, the lectures of Susskind might be of interest.*

Chapter 3

What is a Group?

Finite groups:

A finite group is a set of mathematical objects and a way of combining any two of the objects together which satisfy the group axioms (or have specific properties if the reader prefers the new approach to mathematics). The way of combining any two of the objects together is usually called multiplication. In our case, the mathematical objects will always be square matrices. For finite groups, the way of combining any two of the objects together will always be matrix multiplication[15].

The objects of a finite group are such that:

a) One, and only one, of the objects is the identity. The identity is an object which when multiplied by any other of the objects does not change that second object. In our case, the identity will always by the identity matrix.

$$\begin{bmatrix} 1 & 0 \\ 0 & 1 \end{bmatrix} \quad or \quad \begin{bmatrix} 1 & 0 & 0 \\ 0 & 1 & 0 \\ 0 & 0 & 1 \end{bmatrix} \quad or... \qquad (3.1)$$

---

[15] For continuous groups, the operation will be rotation expressed as multiplication by a rotation matrix.

b) When any two of the objects of a finite group are multiplied together, they produce one of the objects in the finite group. This is multiplicative closure. For example, consider the two matrices:

$$\begin{bmatrix} 1 & 0 \\ 0 & 1 \end{bmatrix} \quad \& \quad \begin{bmatrix} 0 & 1 \\ 1 & 0 \end{bmatrix} \tag{3.2}$$

However we multiply these two matrices together, we always get one or the other of them. These two matrices together are the finite group C_2.

c) The multiplicative inverse of each of the objects of a finite group must be one of the set of objects. For example,

$$\begin{bmatrix} 1 & 0 \\ 0 & 1 \end{bmatrix}^{-1} = \begin{bmatrix} 1 & 0 \\ 0 & 1 \end{bmatrix} \tag{3.3}$$

d) The multiplication operation must be associative (it need not necessarily be commutative). Matrix multiplication is associative, and so, in our case, we will not have to worry about this.

The above four properties are conventionally called the group axioms. Any set of objects which satisfies these axioms is a group.

The order of the group:

A group can consist of any number of objects. The number of objects in the group is called the order of the group. The order can be infinite; for continuous groups, the order is infinite. Finite groups are so called because the number of objects in them is a finite positive whole number. There are finite groups of every positive integer order.

Subgroups:

In many cases, a group of, say, N objects will have within it a sub-set of the N objects which themselves are a group. An example is the finite group $C_2 \times C_2$ which is the four matrices:

$$\begin{bmatrix} 1 & 0 & 0 & 0 \\ 0 & 1 & 0 & 0 \\ 0 & 0 & 1 & 0 \\ 0 & 0 & 0 & 1 \end{bmatrix} \quad \& \quad \begin{bmatrix} 0 & 1 & 0 & 0 \\ 1 & 0 & 0 & 0 \\ 0 & 0 & 0 & 1 \\ 0 & 0 & 1 & 0 \end{bmatrix} \tag{3.4}$$

and:

$$\begin{bmatrix} 0 & 0 & 1 & 0 \\ 0 & 0 & 0 & 1 \\ 1 & 0 & 0 & 0 \\ 0 & 1 & 0 & 0 \end{bmatrix} \quad \& \quad \begin{bmatrix} 0 & 0 & 0 & 1 \\ 0 & 0 & 1 & 0 \\ 0 & 1 & 0 & 0 \\ 1 & 0 & 0 & 0 \end{bmatrix} \tag{3.5}$$

The identity matrix with any one of the other matrices of the group $C_2 \times C_2$ form the group C_2. Thus it is that the group $C_2 \times C_2$ has three C_2 sub-groups within it. Note that the identity matrix must be one of the objects in every sub-group of the group.

Continuous groups:

Continuous groups have the same properties, satisfy the same axioms if the reader prefers, as finite groups.

There are two types of continuous groups corresponding to two types of space. One type of space is a division algebra space which is also called a spinor space. The other type of space is a geometric space which is also called a fabricated space. Our 4-dimensional space-time is a geometric space. You can wave your arms around in a geometric space; other than in 2-dimensional spaces, you cannot wave your arms around in a spinor space. We will be looking at both of these two types of space in more detail shortly.

A continuous group is a set of points in space; each point is an ordered set of real numbers and is a vector. (We are not concerned with spaces that have complex axes which we opine do not really exist.)

If the space is a division algebra space, spinor space, such as quaternion space, then each vector is a spinor, such as a quaternion, written as a matrix. The rotation operation in the division algebra space is by the single rotation matrix (polar form) appropriate to the type of division algebra. The important point is that there is only one rotation matrix in a spinor space; yes, only one – I must remember that.

If the space is a geometric space like our 4-dimensional space-time, then each point in the continuous group is a vector but not a matrix. The rotation operation is not a single rotation matrix but is a set of 2-dimensional rotation matrices. There is one such 2-dimensional rotation matrix for each 2-dimensional plane in the geometric space.

Remember, only one rotation matrix in a spinor space but lots of rotation matrices in a geometric space.

More sub-groups:

Another example of a group with sub-groups is the surface of a sphere in $\mathbb{R}^3$ space. The surface of the sphere is a 2-dimensional continuous group with an infinite number of elements (points on the surface of the sphere). The operation within this continuous group is rotation. Each great circle within the spherical surface is a 1-dimensional continuous group and a sub-group of the 2-dimensional spherical surface.

Consider two great circles upon a spherical surface which intersect. Think of these great circles as the equator and the Greenwich meridian. Each is a continuous group in its own right. Now, we have to change direction, by 90^0 at the points where the great circles intersect, but we can travel by rotation from any point in the pair of circles to any other point in the pair of circles without leaving the circles.

We can think of this pair of great circles itself as a continuous sub-group of the spherical surface. Alternatively, we can think of this pair of great circles as two copies of the same sub-group differing from each other only by a choice of co-ordinates. We will meet something very similar to this when we look at the $SU(2)$ Lie group. In practice, the view we hold of the two great circles seems to be irrelevant although the two separate sub-groups differing by a co-ordinate change view perhaps sits more comfortably in our minds.

Chapter 4

Lie groups in 2-dimensional Spaces

There are two, and only two, 2-dimensional spinor spaces which emerge from the finite groups. These both emerge from the finite group C_2 [16]. These two spinor spaces are each isomorphic to one of the two 2-dimensional geometric spaces which are sub-spaces of the 4-dimensional geometric space which is our 4-dimensional space-time.

The unitary 2-dimensional spinor Lie group:
One of these 2-dimensional spinor spaces is the complex plane, $\mathbb{C}$. Points in this plane are vectors (complex numbers) of the form:

$$\mathbb{C} = \begin{bmatrix} a & b \\ -b & a \end{bmatrix} \tag{4.1}$$

Rotation in this plane is done by the rotation matrix:

$$\exp\left(\begin{bmatrix} 0 & b \\ -b & 0 \end{bmatrix}\right) = \begin{bmatrix} \cos b & \sin b \\ -\sin b & \cos b \end{bmatrix} \tag{4.2}$$

The Lie group of this space is the circle of unit radius in the complex plane. (Although, for convenience, we take the

---

[16] See: Dennis Morris: Complex Numbers The Higher Dimensional Forms.

circle to be of unit radius, we could equally well have chosen any value of the radius; the algebraic and geometric properties are the same regardless of the size of the circle.) We can write this group as the rotation matrix for all values of b. This Lie group is called $U(1)$. We have:

$$U(1)_{Group} = \begin{bmatrix} \cos b & \sin b \\ -\sin b & \cos b \end{bmatrix} \tag{4.3}$$

Note that this space, the complex plane, is a spinor space; it has one real axis and one imaginary axis. A spinor space is a space in which only one axial variable is a real number and in which the other axial variables are imaginary numbers – a spinor space is a type of complex number. Spinor spaces have particular properties:

a) Rotation is not about an axis – there are no constant eigenvectors in the rotation matrix. The complex plane is 2-dimensional; there is no spare dimension to be an axis.

b) Other than in two dimensions, we cannot wave our arms around in spinor spaces. Spinor spaces in general do not have 2-dimensional rotation associated with every possible pair of axial variables – they are not geometric spaces.

c) Spinor spaces cannot be curved. They are globally flat. The particular linear algebra which is the spinor space would fall to bits if the space was not globally flat.

The reader will often see the Lie group $U(1)$ written as:

$$\begin{bmatrix} e^{i\theta} \end{bmatrix} \quad or \quad \begin{bmatrix} e^{-i\theta} \end{bmatrix} \quad or \quad e^{i\theta} \, or \, e^{-i\theta} \tag{4.4}$$

The idea behind the nomenclature is that $U(1)$ is a 1×1 unitary matrix. Of course, whether $U(1)$ is a 1×1 matrix or a 2×2 matrix is just notation, and so the name is uninspiring.

The other 2-dimensional spinor Lie group:

There are two division algebras which emerge from the C_2 finite group. We have dealt with the Euclidean complex numbers, $\mathbb{C}$, above. The other division algebra is the hyperbolic complex numbers, $\mathbb{S}$, which are of the form[17]:

$$\mathbb{S} = \exp\left(\begin{bmatrix} a & b \\ b & a \end{bmatrix}\right) = \begin{bmatrix} e^a & 0 \\ 0 & e^a \end{bmatrix}\begin{bmatrix} \cosh b & \sinh b \\ \sinh b & \cosh b \end{bmatrix} \quad (4.5)$$

The Lie group associated with this division algebra is the hyperbola (one branch only[18]) in the 2-dimensional space-time plane given by the set of points:

$$\begin{bmatrix} \cosh b & \sinh b \\ \sinh b & \cosh b \end{bmatrix} \quad (4.6)$$

This is rotation in 2-dimensional space-time and is often referred to as the Lorentz boost because rotation in 2-

---

[17] Technically, this satisfies the division algebra axioms only if we include a $\pm$ before the radial matrix because we need additive inverses on the real axis to satisfy the algebraic field axioms. Since time flows forwards only, we prefer to do without the $\pm$.

[18] One branch only matches the one direction of time. We cannot travel backwards in time, and so we take only the positive branch of the hyperbola.

dimensional space-time is a change of velocity. This too is a spinor space. By the way, a little trigonometry will lead to[19]:

$$\cosh \chi = \frac{1}{\sqrt{1 - \frac{v^2}{c^2}}}, \qquad \sinh \chi = \frac{v}{\sqrt{1 - \frac{v^2}{c^2}}} \qquad (4.7)$$

This 2-dimensional Lie group, (4.6), as far as your author is aware, has no name or nomenclature. It should not be confused with the 4-dimensional Lorentz group, $SO(3,1)$ which is the Lie group of a geometric space. Lie groups in spinor spaces are very different things from Lie groups in geometric spaces.

Compact and non-compact:

The circle is such that we can rotate all the way around the origin. Such a Lie group is said to be compact. The hyperbola is such that its two ends never meet. Such a Lie group is called non-compact.

Geometric 2-dimensional spaces:

The reader might have noticed that a 2-dimensional being could, if he existed, wave his arms around in all possible directions in either of the two 2-dimensional spinor spaces $\mathbb{C} \& \mathbb{S}$. This is because there are no 'extra' dimensions to form 2-dimensional planes other than the single 2-

---

[19] See : Empty Space is Amazing Stuff by Dennis Morris. Your author apologises to the reader for so often referring to his own books, but these are the references best known to him.

dimensional plane which is the spinor space. If you can wave your arms around in a space, then the space is a geometric space. The 2-dimensional spaces are special in that they are both geometric spaces at the same time as being spinor spaces. There are no other spinor spaces that are also geometric spaces.

To be more technical, the geometric Lie group, $SO(3,1)$, of our 4-dimensional $\mathbb{R}^4$ space-time has two types of 2-dimensional geometric sub-groups in $\mathbb{R}^2$ which are Euclidean rotation and hyperbolic rotation. These geometric sub-groups might be written as $SO(2)$ and $SO(1,1)$, but the $SO(1,1)$ is not used because the hyperbolic rotation matrix is not an orthogonal matrix.

Generators:
The Lie algebra associated with a Lie group is a set of matrices which, when multiplied by a real variable, have exponentials that give the Lie group – more on this later. The generators of the two 2-dimensional Lie groups are:

$$U(1) \sim \begin{bmatrix} 0 & 1 \\ -1 & 0 \end{bmatrix} \qquad L(1) \sim \begin{bmatrix} 0 & 1 \\ 1 & 0 \end{bmatrix} \qquad (4.8)$$

We have used the nomenclature $L(1)$ to refer to the Lie group in 2-dimensional space-time. Multiplied by a variable, these, (4.8), become:

$$U(1) \sim \begin{bmatrix} 0 & \theta \\ -\theta & 0 \end{bmatrix} \qquad L(1) \sim \begin{bmatrix} 0 & \chi \\ \chi & 0 \end{bmatrix} \qquad (4.9)$$

Taking the exponential of each of these gives one or other of the rotation matrices, Lie groups in the 2-dimensional case, (4.3) & (4.6).

The reader might often see these generators written in different bases or even written as a single real number, θ, with the understanding that we are to take the exponential of $i\theta$ to form the Lie group.

In general, every 2-dimensional circular rotation has a generator equal, except in direction, to $U(1)$ and every 2-dimensional hyperbolic rotation has a generator equal, except in direction, to $L(1)$. Sometimes, this simplicity is confused by multiplying the generators by a number like $i = \sqrt[2]{-1}$.

Summary of 2-dim spinor Lie groups:
There are two 2-dimensional spinor Lie groups because there are two types of 2-dimensional spinor space – two types of 2-dimensional complex numbers. One Lie group is the circle; the other Lie group is the hyperbola (one branch only). These are both division algebra Lie groups and geometric space Lie groups at the same time.

Summary of 2-dim geometric Lie groups:
There are two 2-dimensional geometric Lie groups because there are two types of 2-dimensional $\mathbb{R}^2$ space. One Lie group is the circle; the other Lie group is the hyperbola (one branch only). These are both division algebra Lie groups and geometric space Lie groups at the same time.

Chapter 5

Lie groups in 3-dimensional space

There are two types of Lie groups in 3-dimensional space. There are four spinor Lie groups within the finite group C_3 (two are isomorphic), and there is the rotational surface of a sphere in the geometric $\mathbb{R}^3$ space which we call $SO(3)$. The Lie group $SO(3)$ is a sub-group of the Lorentz group $SO(3,1)$ of our 4-dimensional space-time.

The geometric $\mathbb{R}^3$ space does not exist in its own right; it exists only within the geometric 4-dimensional space-time of our universe as a sub-space – we never see $\mathbb{R}^3$ space without time. Thus it is that $SO(3)$ does not exist in its own right but only as a sub-group of the Lorentz group. None-the-less, $SO(3)$ is often taken to exist in its own right, and no harm comes from doing that, and so we will do the same. We will not deal with $SO(3)$ in this chapter because it is more properly dealt with when we look at our 4-dimensional space-time. In this chapter, we deal with only the spinor Lie groups within the finite group C_3.

3-dimensional spinor Lie groups:
The finite group C_3 holds four 3-dimensional division algebras (types of complex numbers). Each of these algebras

has a polar form. The rotation matrix in each of the polar forms of these algebras gives a continuous set of points of unit distance, as measured with the appropriate distance function, from the origin, and so there are four 3-dimensional Lie groups. Each of the four 3-dimensional spinor Lie groups is the rotation matrix of the polar form of each of the four spinor algebras. Two of the 3-dimensional division algebras are algebraically isomorphic, and so their Lie groups are isomorphic as Lie groups.

These 3-dimensional Lie groups are not traditionally considered to be Lie groups because they were unknown until 2007[20]. Unless the reader has seen this before, the reader is about to see something weird. Actually, even if the reader has seen this before, it is still weird.

The algebraic matrix form of one of the four 3-dimensional division algebras, spinor spaces, is:

$$\exp\left(\begin{bmatrix} a & b & c \\ c & a & b \\ b & c & a \end{bmatrix}\right) \qquad (5.1)$$

The algebraic matrix forms of the other three 3-dimensional spinor spaces are the same as (5.1) but with three different scatterings of minus signs around the matrix.

The Lie group is the rotation matrix in this 3-dimensional spinor space, division algebra space, as:

[20] See: Dennis Morris: Complex Numbers the Higher Dimensional Forms.

$$\exp\left(\begin{bmatrix} 0 & b & c \\ c & 0 & b \\ b & c & 0 \end{bmatrix}\right) = \begin{bmatrix} v_A & v_B & v_C \\ v_C & v_A & v_B \\ v_B & v_C & v_A \end{bmatrix} \qquad (5.2)$$

The matrix with the nu-functions, v_i, is the rotation matrix of this space and hence is the set of points which is the Lie group, continuous group, of this space. The nu-functions are the 3-dimensional trigonometric functions. The details of the nu-functions are given elsewhere[21]. This rotation matrix respects the distance function (which is the determinant of the matrix inside the exponential in (5.1)):

$$dist^3 = a^3 + b^3 + c^3 - 3abc \qquad (5.3)$$

The interesting point about this Lie group is that, because the finite group C_3 has no C_2 sub-groups, there are no 2-dimensional rotations in this 3-dimensional space.

It is not hard to show that the rotation matrix satisfies all the group axioms, inverse and that kind of stuff, and so this is definitely a continuous group even though tradition excludes it from being called a Lie group. This Lie group is non-compact.

Within this division algebra space, we have the standard characteristics of a spinor space; there is no rotation about an axis, the space is flat, and we cannot wave our arms around in this space. Why can we not wave our arms around in this space? There are three axial variables corresponding to the three axes of this space. These three variables are all

---

[21] See: Dennis Morris: Complex Numbers the Higher Dimensional Forms.

connected together in a single 3-dimensional rotation, but there are no 2-dimensional rotations in 2-dimensional planes within this space because there are no 2-dimensional planes in this space. Arguably, we can wave our arms 3-dimensionally as we waved them 2-dimensionally in the 2-dimensional spinor spaces, but the essence is that this is not a geometric space like our 4-dimensional space-time.

This 3-dimensional division algebra is of the form:

$$a + b\sqrt[3]{+1} + c\sqrt[3]{+1} \qquad (5.4)$$

We have cube roots for the imaginary variables. The other three division algebras are of the forms:

$$\begin{aligned} a + b\sqrt[3]{+1} + c\sqrt[3]{-1} \\ a + b\sqrt[3]{-1} + c\sqrt[3]{+1} \\ a + b\sqrt[3]{-1} + c\sqrt[3]{-1} \end{aligned} \qquad (5.5)$$

The generators of the Lie algebra associated with this Lie group, (5.1), would conventionally be presented as[22]:

$$\begin{bmatrix} 0 & 1 & 0 \\ 0 & 0 & 1 \\ 1 & 0 & 0 \end{bmatrix} \quad \& \quad \begin{bmatrix} 0 & 0 & 1 \\ 1 & 0 & 0 \\ 0 & 1 & 0 \end{bmatrix} \qquad (5.6)$$

However, there is an error here. The conventional presentation of generators assumes rotation in separate (2-dimensional) planes; there are no separate planes in this

---

[22] More on generators later. Roughly, we take the exponential of the generators to get the rotation matrix.

spinor space. More sensibly, we should present the generator as:

$$\begin{bmatrix} 0 & 1_b & 1_c \\ 1_c & 0 & 1_b \\ 1_b & 1_c & 0 \end{bmatrix} \tag{5.7}$$

Because this is a commutative spinor space, it matters not whether we take the exponential of (5.7) or take the exponentials of both (5.6) and then combine the two together.

We need to keep in mind that the Lie group is the important part of all this.

A pretend game:

Let us pretend that the group C_3 has two C_2 sub-groups. Those two C_2 sub-groups would be comprised of the identity (the real number) and one of the imaginary numbers. These would then be 2-dimensional division algebras, and we could rotate within these 2-dimensional sub-spaces. We would not be able to form a C_2 sub-group from only the two imaginary numbers because a group, or sub-group, must include the identity, the real number. In our pretend space, we would be able to rotate in only two of the three 2-dimensional planes. This is insufficient to allow us to wave our arms around. The message of this pretend game will become of great importance when we look at the 4-dimensional quaternion space.

There is another point. We see that the 3-dimensional rotation matrix, (5.2), has only two angle parameters. To rotate in three 2-dimensional planes requires three angle parameters. We have no third angle parameter to measure rotation in a third 2-dimensional plane.

3-dim hand waving – more pretend games:

Suppose there was a 4-dimensional space with the distance function:

$$dist^3 = a^3 + b^3 + c^3 + d^3 - 3abc - 3abd - 3acd - 3bcd \quad (5.8)$$

In such a space, if we had the requisite number of angle parameters, we could wave our arms 3-dimensionally around in four different 3-dimensional 'planes'. There is no triple of variables which does not support 3-dimensional rotation – respect the 3-dimensional distance function.

If such a 4-dimensional space were curved, that curvature would be defined by four pairs of real numbers which we would call the principle curvatures of the space. Such a space would be a geometric space; it would be something very different from the geometric space which is our 4-dimensional space-time with its 2-dimensional rotations, but it would be a space in which one could wave one's arms 3-dimensionally.

Such a space does not exist within the new approach mathematics because it does not derive from the finite groups. Of course, if he so wishes an axiomatic approach mathematician can create such a space in his imagination, add in four pairs of real parameters as the four 3-

dimensional angles, and, bingo, we have a type of space that does not exist.

Summary:

We used the 3-dimensional division algebras, 3-dimensional spinors, to air the ideas which the reader will need when confronting quaternion space. Those ideas are:

a) There are rotations which are not 2-dimensional rotations. Every spinor space, has rotations of dimension equal to the order of the finite group which underlies that spinor space. We saw 3-dimensional rotations.

b) Not every space has 2-dimensional planes within it.

c) Even spinor spaces which have 2-dimensional planes within them do not have these planes between all pairs of variables – you cannot form a 2-dimensional plane in spinor space between two imaginary variables.

d) A geometric space of 2-dimensional rotations needs as many angle parameters as there are pairs of variables. Except in the 2-dimensional cases, spinor spaces do not have the number of angle parameters required to allow geometric rotation – also known as arm waving.

e) Not all distance functions respected by rotations are of the Riemann nature.

Chapter 6

Understanding Georgi

In this chapter, we introduce the conventional approach to Lie theory. We do this because the student will need to understand the conventional approach to read the literature and because we want to make clear the view of empty space adopted by conventional Lie theory. Having read the previous chapters, the limitations of the conventional view of empty space become lucent.

The standard mantra in Lie algebras and Lie groups is well expressed in the book by Howard Georgi entitled 'Lie Algebras in Particle Physics'. This book is very respected and is rapidly becoming a classic. We quote extensively from Georgi's book in this chapter.

Georgi chapter two – Lie Groups:
In chapter two[23], page 43, of Georgi's book, Georgi formulates a Lie group as a set of group elements, $g \in G$, which each depend upon a single parameter, $g(\alpha)$[24]. It becomes apparent later in the book and from a general

---

[23] Chapter two of Georgi begins on page 43 in my edition of the book.
[24] This is equation (2.1) in Georgi.

41

understanding of Lie groups that Georgi is really talking about rotations.

The only type of rotations which depend upon a single parameter are the 2-dimensional spinor rotations. 3-dimensional spinor rotations depend upon two parameters, and 4-dimensional rotations depend on three parameters etc.. Convention might have considered formulating a Lie group as a single element which depends on a number of parameters, $G(\alpha, \beta, \gamma...)$, but it did not. We see here in Georgi that Georgi is formulating a Lie group as being in a space in which there are a number of separate rotations, one for each element of the group, and in which every rotation is a 2-dimensional rotation depending upon a single parameter which is a single angle. In this book, we refer to such spaces as geometric spaces and we accept only the geometric spaces that derive from the finite groups. Georgi differs from this book in that he does not require the geometric space to be derived from the finite groups but will allow any construction which is consistent with the axioms of mathematics.

Georgi concurs with this 'rotation' view. In his book, on page 240, 9.1, we have, "*Obviously SO(N) is the group of rotations in real N-dimensional space.*" Your author agrees with that. Later in the same paragraph, we also have, "*Similarly, SU(N) is the group of "rotations" in complex N-dimensional space...*" Your author does not believe in complex axes and opines that Georgi has got this bit wrong[25]. In the next paragraph, Georgi gives us, "*Finally,*

---

[25] Which shows that even very brainy people can make mistakes, in your author's opinion.

$Sp(2N)$ *can be thought of as the group of rotations in a quaternion N-dimensional space...*" Your author does not believe in quaternion axes. None-the-less, Georgi has the rotation view in heart. Let us return to page 43 and continue.

Later, on page 43, Georgi gets the identity by setting the angle parameter to zero, $\alpha = 0$. Georgi remains in the abstract linear transformation mind until he adopts a representation. A representation is a matrix; we get:

$$D(\alpha)\big|_{\alpha=0} = 1 \qquad (6.1)$$

This is just a 2-dimensional rotation:

$$\begin{bmatrix} \cos\alpha & \sin\alpha \\ -\sin\alpha & \cos\alpha \end{bmatrix}\bigg|_{\alpha=0} = \begin{bmatrix} 1 & 0 \\ 0 & 1 \end{bmatrix} \qquad (6.2)$$

There is an aside which Georgi does not mention. Georgi's concern is with Euclidean rotations, but (6.2) could have been a hyperbolic rotation:

$$\begin{bmatrix} \cosh\alpha & \sinh\alpha \\ \sinh\alpha & \cosh\alpha \end{bmatrix}\bigg|_{\alpha=0} = \begin{bmatrix} 1 & 0 \\ 0 & 1 \end{bmatrix} \qquad (6.3)$$

Having mentioned this aside, we will stick with the Euclidean version for ease.

Georgi's next step is to Taylor expand the representation, (6.1), for very small α. Because $\alpha = d\alpha$ is infinitesimally small, Georgi can ignore the higher order terms in the Taylor expansion and is thus led to define a generator of the Lie group as[26]:

$$X_a \equiv -i \frac{\partial D(\alpha)}{\partial \alpha_a}\Big|_{\alpha=0} \tag{6.4}$$

Of course, this applies to each generator; hence the subscripted a. We have:

$$\frac{\partial \begin{bmatrix} \cos\alpha_a & \sin\alpha_a \\ -\sin\alpha_a & \cos\alpha_a \end{bmatrix}}{\partial \begin{bmatrix} 0 & \alpha_a \\ -\alpha_a & 0 \end{bmatrix}} = \begin{bmatrix} 0 & -1 \\ 1 & 0 \end{bmatrix} \begin{bmatrix} -\sin\alpha_a & \cos\alpha_a \\ -\cos\alpha_a & -\sin\alpha_a \end{bmatrix}$$

$$= \begin{bmatrix} \cos\alpha_a & \sin\alpha_a \\ -\sin\alpha_a & \cos\alpha_a \end{bmatrix} \tag{6.5}$$

Multiplying this by $-i$ and evaluating at $\alpha = 0$ gives:

$$\begin{bmatrix} 0 & -1 \\ 1 & 0 \end{bmatrix} \begin{bmatrix} \cos\alpha_a & \sin\alpha_a \\ -\sin\alpha_a & \cos\alpha_a \end{bmatrix}\Big|_{\alpha=0} = \begin{bmatrix} \sin\alpha_a & -\cos\alpha_a \\ \cos\alpha_a & \sin\alpha_a \end{bmatrix}\Big|_{\alpha=0}$$

$$= \begin{bmatrix} 0 & -1 \\ 1 & 0 \end{bmatrix} \tag{6.6}$$

Which is the generator of a 2-dimensional Euclidean rotation. The 2-dimensional Euclidean rotation is generated by taking the exponential of this generator matrix. (We

---

[26] This is equation (2.5) in Georgi.

could have done something similar with the hyperbolic rotation.)

We see the standard mantra as expressed by Georgi completely excludes the spinor rotations (other than in 2-dimensions). This standard mantra is based upon the erroneous assumption that all rotations are 2-dimensional and that all spaces are such that they have only 2-dimensional rotations within them and that there are a number of such 2-dimensional rotations which are orthogonal to each other.

We ought not to contemn convention, and particularly Georgi, too harshly. We have the advantage of using matrix notation; without matrix notation, it is easy to think all rotations are 2-dimensional. For example, quaternions have three 2-dimensional sub-algebras. In matrix notation, as we will see, these 2-dimensional sub-algebras are 4×4 matrices which double cover the 2-dimensional complex numbers written as a 2×2 matrix. In matrix notation, we can see that the 4×4 2-dimensional sub-algebras are not the same as the 2×2 complex numbers. In conventional notation, a quaternion is written as $a + ib + jc + kd$. If two of the imaginary variables are set to zero we get $a + ib$, and this looks very much like a complex number written as $a + ib$. No wonder people used to think that quaternions have complex numbers for sub-algebras, and three of them; it is not quite true as will be explained later. Thus, we have, in quaternion space, a space, or so the conventional view goes, with three 2-dimensional rotations. It is easy to see why the 'only 2-dimensional rotations' view was adopted by convention. While we are at it, we might as well point out that there are 8-dimensional division algebras which have seven 2-dimensional sub-algebras within them.

That Georgi is really describing sets of 2-dimensional rotations is re-emphasized[27] by the passage:

"..in any particular direction… the group multiplication law is…

$$U\left(\lambda_1\right)U\left(\lambda_2\right)=U\left(\lambda_1+\lambda_2\right)"\tag{6.7}$$

This is clearly two successive rotations.

The commutation algebra:
Georgi goes on to deal with non-commutativity[28]:

$$e^{i\alpha_1 X_1}e^{i\alpha_2 X_2}\neq e^{i\left(\alpha_1 X_1+\alpha_2 X_2\right)}\tag{6.8}$$

How do you deal with such non-commutativity? Following the conventional exposition, Georgi assumes that the product of the two non-commutative exponentials, whatever it be, is within the group. Intuitively, within a geometric space, two successive rotations will 'end' at another point in the rotational surface. Georgi's assumption is predicated upon the space being of a geometrical nature with such a rotational surface. Georgi continues with, "… *it only works if the generators form an algebra under commutation (or a commutator algebra).*" We have it that the product of the two exponentials is within the group only if the generators form a commutator algebra.

---

[27] Georgi page 45 equation (2.8)
[28] Page 45 equation (2.10).

The requirement for the commutator algebra now makes it very clear that we are not dealing with a geometric rotational surface like the surface of a sphere in $\mathbb{R}^3$ space. Any two rotations in $\mathbb{R}^3$ space are such that their composition ends at another point in the rotational surface; there are no conditions upon this result in $\mathbb{R}^3$ space; we do not require any generators to form a commutation algebra in $\mathbb{R}^3$ space.

This book (the one you are reading, not Georgi's book) now has things a little out of sequence in that we need to briefly introduce the reader to the concept of a commutation algebra and we need to draw on results presented later in this book. Perhaps the next few paragraphs will make more sense to the reader if the reader returns to these few paragraphs after reading the rest of this book.

A commutator algebra is a Lie algebra. There are two types of Lie algebra because there are two types of Lie group. Spinor Lie groups such as the quaternions form a commutation algebra of the form:

$$\left[\hat{i}, \quad j \right] = \hat{i}j - j\hat{i} = 2k \tag{6.9}$$

Geometric Lie groups also have commutation algebras like[29]:

$$\left[X_a, X_b \right] = X_a X_b - X_b X_a = i f_{abc} X_c \tag{6.10}$$

The expression in the brackets is known as the commutator which we now consider.

---

[29] We have pinched this expression from Georgi pg 46.

As will be clearer after the reader has read the chapters on the finite group S_3 commutation relations and on the 8-dimensional Lie groups, we can have a commutator algebra of the required form only if the generators form a non-commutative 4-dimensional algebra from the $C_2 \times C_2$ finite group.

Putting this slightly differently, the product of the two exponentials is within the group only if the generators form a non-commutative 4-dimensional algebra from the $C_2 \times C_2$ finite group (or its expectation spaces).

In later chapters, and where previously stated, we see that the Lie group $SU(2)$ is a spinor Lie group consisting of three intersecting circles and is not a rotational surface within a geometric space. We are now able to understand why the product of the two exponentials is within the group only if the generators form a commutator algebra. The product of the two exponentials is within the group only if the generators are on one of the three intersecting circles which is the spinor Lie group.

Georgi goes on to say in astonishment, "*The commutator relation is enough.*"[30] Of course the commutator relation is enough; every element of the spinor Lie group is within a multiplicatively closed division algebra; any product of any elements within that multiplicatively closed algebra will be within that algebra because it is multiplicatively closed.

The central aspect of the conventional formation of a Lie algebra is the "... *it only works if the generators form an*

---

[30] Page 47.

algebra under commutation (or a commutator algebra)." and the *"The commutator relation is enough."* The conventional approach is reaching toward the spinor quaternion space, grasping the spinor quaternion space, and not really knowing anything about the spinor quaternion spaces which it now holds.

Confusion upon confusion:

The reason Lie algebra and Lie groups seem so confusing to the student is that they are confusing. The whole of conventional Lie algebra and Lie group theory is based upon the assumption that there are only 2-dimensional rotations. The higher dimensional spinor rotations are ignored because they were unknown at the time Lie theory was formulated. This view was enforced by the fact that the very important Lie group $SU(2)$ is a 4-dimensional rotation group which seems to hold 2-dimensional rotations. Given that the mathematicians who know Lie theory have it wrong, it is hardly surprising that the student gets confused.

The standard mantra, not Georgi, proceeds by saying, "We need generators which preserve the 2-dimensional distance function $dist^2 = x^2 + y^2$." Such generators are obviously 2-Euclidean dimensional rotations in some form. All unitary matrices preserve this 2-dimensional distance function. Hence the standard mantra is to collect together all possible sets of unitary matrices of the same size. We have to be careful to ensure the unitary matrices are not actually duplicates written in different bases, but we can do that. Hence, we might as well call these sets of unitary matrices the unitary Lie algebras and give them names like $U(1)$ or

$SU(2)$ wherein the number is the size of the matrices with complex elements. Using complex elements in the matrices rather than real elements does not help – it does save paper.

The next step is to forget all about rotations and distance functions and assert that the distance expression is really a probability $\text{Prob} = x^2 + y^2$, and so we now have rotations in an abstract probability space. Even more confusing, since we are using complex numbers in our unitary matrices, we assert the probability space has complex axes – it's a complex Hilbert space. By such means does misunderstanding breed incestuously with itself and grow large and overpowering. Complex axes! are we stark raving bonkers?

Feeding upon the idea that higher dimensional spinor rotations do not exist[31], quantum physics has taken the best that was available and built a great edifice called the standard model based upon the unitary Lie groups $U(1)$, $SU(2)$, and $SU(3)$. It is amazing that physicists have achieved this given the apparent non-existence of $SU(3)$ upon which we will write later in this book. Within this book, we will shortly give the reader much insight into $SU(2)$ and we will see that the physicists have this bit correct. We do not know what to do about the wayward $SU(3)$, and so we will have to leave it until we understand how to replace it.

---

[31] Spinor rotations were unknown until 2007 and it was five years after that before the non-commutative rotations were known, and so there is room for sympathy.

Summary:

We hope that within this chapter we have given the reader confidence in our assertion that conventional Lie theory is predicated upon the only types of space being geometric spaces containing only 2-dimensional rotations. The chapters following this one will develop and expand our understanding of Lie groups as rotational surfaces within both the spinor spaces and our 4-dimensional space-time. We will deal with Lie algebras later.

Chapter 7

4-dimensional Lie Groups – Part 1

There are two order four finite groups. These are the C_4 group and the $C_2 \times C_2$ group. Each of these groups holds spinor spaces and thus contains spinor Lie groups. As well as the spinor Lie groups, there is also a 4-dimensional Lie group known as the Lorentz group, $SO(3,1)$ associated with our 4-dimensional space-time and a 4-dimensional Lie group known as $SO(4)$. These are Lie groups in geometric spaces. This chapter is concerned with spinor Lie groups, and so we will not deal with $SO(3,1)$ or $SO(4)$ in this chapter.

C_4 space:

The group C_4 has one order two sub-group. Thus, the 4-dimensional C_4 space has a 2-dimensional sub-space in only one 2-dimensional plane. There is only one 2-dimensional plane in the 4-dimensional C_4 space. Yes, it blows your mind a little does it not? Clearly, this is a different kind of space from that to which we are accustomed.

There are eight division algebras, spinor algebras, in the C_4 group, but they are two sets of four isomorphic algebras. Thus, there are two types of rotation in the C_4 group.

The rotation matrices are the hyperbolic type:

$$PROD\left\{\begin{bmatrix} AH_4(b) & BH_4(b) & CH_4(b) & DH_4(b) \\ DH_4(b) & AH_4(b) & BH_4(b) & CH_4(b) \\ CH_4(b) & DH_4(b) & AH_4(b) & BH_4(b) \\ BH_4(b) & CH_4(b) & DH_4(b) & AH_4(b) \end{bmatrix}\begin{bmatrix} \cosh c & 0 & \sinh c & 0 \\ 0 & \cosh c & 0 & \sinh c \\ \sinh c & 0 & \cosh c & 0 \\ 0 & \sinh c & 0 & \cosh c \end{bmatrix}\begin{bmatrix} AH_4(d) & DH_4(d) & CH_4(d) & BH_4(d) \\ BH_4(d) & AH_4(d) & DH_4(d) & CH_4(d) \\ CH_4(d) & BH_4(d) & AH_4(d) & DH_4(d) \\ DH_4(d) & CH_4(d) & BH_4(d) & AH_4(d) \end{bmatrix}\right\}$$

(7.1)

$$dist^4 = \left((a+c)^2 - (b+d)^2\right)\left((a-c)^2 + (b-d)^2\right)$$

(7.2)

And the Euclidean type:

$$PROD\left\{\begin{array}{l} \begin{bmatrix} AE_4(b) & BE_4(b) & CE_4(b) & DE_4(b) \\ -DE_4(b) & AE_4(b) & BE_4(b) & -CE_4(b) \\ -CE_4(b) & -DE_4(b) & AE_4(b) & -BE_4(b) \\ BE_4(b) & CE_4(b) & DE_4(b) & AE_4(b) \end{bmatrix} \\ \begin{bmatrix} \cos c & 0 & \sin c & 0 \\ 0 & \cos c & 0 & \sin c \\ \sin c & 0 & \cos c & 0 \\ 0 & \sin c & 0 & \cos c \end{bmatrix} \\ \begin{bmatrix} AE_4(d) & -DE_4(d) & -CE_4(d) & BE_4(d) \\ BE_4(d) & AE_4(d) & -DE_4(d) & CE_4(d) \\ CE_4(d) & BE_4(d) & AE_4(d) & DE_4(d) \\ -DE_4(d) & -CE_4(d) & -BE_4(d) & AE_4(d) \end{bmatrix} \end{array}\right\}$$

$$(7.3)$$

$$dist^4 = \left(a^2 + c^2\right)^2 + \left(b^2 + d^2\right)^2$$
$$+ 4\left(\left(a^2 - c^2\right)bd + \left(d^2 - b^2\right)ac\right)$$

$$(7.4)$$

The details of the C_4 simple trigonometric functions, AH_4... and AE_4... are given elsewhere[32]. These rotation matrices respect the distance functions shown. The pairing together of variables seen in the distance functions is typical of spinor spaces within finite groups which hold sub-groups within them.

---

[32] See: Dennis Morris: Complex Numbers The Higher Dimensional Forms.

We are not greatly interested in the Lie groups of the C_4 group for it seems they play no role in our physical universe. We looked at the C_4 group only for completeness and to familiarise the reader with the 'weird' nature of a space with 2-dimensional planes linking only some of the possible pairs of variables.

The $C_2 \times C_2$ Lie groups:

The finite group $C_2 \times C_2$ contains the whole of classical physics, seemingly much of quantum physics and, perhaps, the whole of quantum physics. We are therefore very interested in its Lie groups. We will take a preliminary look at the $C_2 \times C_2$ Lie groups in this chapter. We will then look at the Lie groups of our 4-dimensional space-time in the next chapter. We will then come back to the $C_2 \times C_2$ Lie groups in the chapter following the 'our 4-dimensional space-time' chapter.

It is the nature of the $C_2 \times C_2 \times ...$ groups in general that they have lots of 2-dimensional sub-groups. In fact, in the spinor algebras of these groups, every imaginary variable, together with the real variable, forms a 2-dimensional sub-algebra. This makes them 'similar' in a very loose sense to a geometric space.

The $C_2 \times C_2$ group contains sixteen separate division algebras of four non-isomorphic types. Remarkably, eight of the sixteen division algebras of this commutative group are non-commutative division algebras. These non-commutative algebras are our primary interest, but we will look at the commutative algebras first.

The commutative $C_2 \times C_2$ Lie groups:

There are two commutative $C_2 \times C_2$ division algebras comprised of one real variable and three symmetric imaginary variables. We call these the A_1 algebras.

A symmetric variable is one which appears symmetrically across the leading diagonal in the algebraic matrix form of the algebra. A symmetric variable is a square root of plus unity; an anti-symmetric variable is a square root of minus unity.

There are six commutative $C_2 \times C_2$ division algebras comprised of one real variable and one symmetric imaginary variable and two anti-symmetric imaginary variables. We call these the A_2 algebras.

Aside: Symmetric matrices have real eigenvalues. Anti-symmetric matrices have only imaginary eigenvalues (and zero if the matrix is of odd size) which are in conjugate pairs.

The A_1 algebras:

The two $C_2 \times C_2$ algebras comprised of one real variable and three symmetric imaginary variables are isomorphic and differ only by the basis in which they are written. They have the form:

$$\exp\left(\begin{bmatrix} a & b & c & d \\ b & a & d & c \\ c & d & a & b \\ d & c & b & a \end{bmatrix}\right) \tag{7.5}$$

$$a + b\sqrt[2]{+1} + c\sqrt[2]{+1} + d\sqrt[2]{+1} \tag{7.6}$$

The A_1 Lie groups are of the form:

$$PROD\left\{\begin{bmatrix} \cosh b & \sinh b & 0 & 0 \\ \sinh b & \cosh b & 0 & 0 \\ 0 & 0 & \cosh b & \sinh b \\ 0 & 0 & \sinh b & \cosh b \end{bmatrix} \begin{bmatrix} \cosh c & 0 & \sinh c & 0 \\ 0 & \cosh c & 0 & \sinh c \\ \sinh c & 0 & \cosh c & 0 \\ 0 & \sinh c & 0 & \cosh c \end{bmatrix} \begin{bmatrix} \cosh d & 0 & 0 & \sinh d \\ 0 & \cosh d & \sinh d & 0 \\ 0 & \sinh d & \cosh d & 0 \\ \sinh d & 0 & 0 & \cosh d \end{bmatrix}\right\} \tag{7.7}$$

This form reflects the fact that the $C_2 \times C_2$ group has three C_2 sub-groups and that, in the case of these two algebras, the imaginary variables are symmetric within the algebraic matrix form.

The above (7.7) is not rotation about an axis. There is no 'spare' dimension to form an axis. This is spinor rotation with no axis, as taking the eigenvectors will confirm. Of course, all Lie groups within a division algebra space, a spinor space, are not rotational surfaces about an axis.

These two A_1 Lie groups respect distance functions of the form:

$$dist^4 = \left((a+b)^2 - (c+d)^2\right)\left((a-b)^2 - (c-d)^2\right) \quad (7.8)$$

This will not reduce to a quadratic form, $dist^2 = ...$.

Pairing of variables:
We see the pairing of variables again. Perhaps a little explanation is warranted. There are only two possible 2-dimensional distance functions (2-dimensional division algebras) corresponding to the Euclidean complex numbers, $\mathbb{C}$, and the hyperbolic complex numbers, $\mathbb{S}$. Each of the three 2-dimensional sub-algebras of the 4-dimensional A_1 algebras must have one or other of the 2-dimensional distance functions; in the A_1 case, the sub-algebras are all the hyperbolic complex numbers. What kind of 4-dimensional distance function will, when two variables are set to zero, reduce, in all three cases, to one of the two 2-dimensional distance functions? There are some distance functions that will do this, but just any old function will not do this. In short, the 4-dimensional distance function is

constrained by the need to reduce to a 2-dimensional distance function for each of the three order two sub-groups.

Note: The 8-dimensional $C_2 \times C_2 \times C_2$ algebras have seven order two sub-algebras and seven order four sub-algebras all of which must be accommodated within the 8-dimensional distance functions of these algebras. Of course, the trigonometric functions of an algebra also have to accommodate the trigonometric functions of the sub-algebras.

Taking the distance function to bits:

There are six different ways to pair together two of four variables; that is, there are six possible 2-dimensional planes in 4-dimensional space. Let us take pairs from the above A_1 distance function. We find that, with two variables set to zero, the distance function will reduce to quadratic form. The three sub-algebras are:

$$
\begin{aligned}
dist^2 &= a^2 - b^2 \\
dist^2 &= a^2 - c^2 \\
dist^2 &= a^2 - d^2
\end{aligned}
\tag{7.9}
$$

The three pairings of imaginary variables are:

$$
\begin{aligned}
dist^2 &= b^2 - c^2 \\
dist^2 &= b^2 - d^2 \\
dist^2 &= c^2 - d^2
\end{aligned}
\tag{7.10}
$$

Looking at (7.9) & (7.10), it would seem that we have 2-dimensional rotations in all six 2-dimensional planes. If this were the case, we would have six parameters, angles, associated with this algebra – one for each rotational plane. We have only three parameters, $\{a, b, c\}$ in (7.7). With only three parameters, we can have no more than three 2-dimensional planes.

There is something more 'wrong' with this space. Looking at (7.9) & (7.10), we see that all six potential 2-dimensional rotations are space-time rotations (minus sign). In 2-dimensional space-time rotations, the signs before the squares of the two variables are different (a minus and a plus). It is not possible to have a 3-dimensional or higher dimensional space in which the squares of all three variables have different signs before them because there are only two types of sign, plus or minus. We cannot form any geometric space of dimension more than two which has only 2-dimensional space-time rotations in each 2-dimensional plane.

What does it mean?:

We see that we can rotate 2-dimensionally in only three 2-dimensional planes within the A_1 space. (This applies also to all other spinor algebras which derive from the $C_2 \times C_2$ group.) This space has only three 2-dimensional planes.

Now, pretend we have a 3-dimensional Euclidean space in which we can rotate in only two of the three 2-dimensional planes. Pretend we can rotate in, say, both the orthogonal vertical planes but not in the horizontal plane. Can we rotate in both the vertical planes at the same time? No, because to rotate in both vertical planes at the same time is effectively to partially rotate in the horizontal plane which cannot be done.

We see that, in our pretend space, the rotational surface is just two vertical circles which intersect at two points; this is our pretend Lie group. Rotation in such a space would be very much like the spin of the electron; it would be either spin up or spin down in two possible directions.

Going back to reality and the three 2-dimensional rotational planes in the A_1 spinor space, we see that this Lie group is just three 'hyperbolic circles', hyperbolas, which intersect each other at two points each. Rotation in this space is all 2-dimensional space-time rotation (we cannot rotate all the way around to where we started from); this is not like the spin of the electron except for the rotation being in three discreet planes.

These rotations are also unlike electron spin in that they are commutative – this is a commutative algebra.

A word of warning: The A_1 Lie group looks like a set of intersecting hyperbolas to we who sit in our geometric space. If we were sitting in A_1 space, The Lie group would appear, I presume[33], as a continuous rotational surface.

---

[33] Your author has never been into A_1 space, and so he does not know that of which he writes.

Expectation spaces:

We will see later, as can be seen in more detail elsewhere[34], that our 4-dimensional space-time emerges from the $C_2 \times C_2$ group as an expectation space formed from adding isomorphic spinor spaces.

Perhaps we could add the two A_1 algebras together in some way and thus have the six parameters, angles, we need to form a geometric space. After all is said and done, they are the same algebra just written in two different bases, and physicists do this kind of thing every day when they take expectation values. If we add the two A_1 algebras in some way, then we would have to add the two distance functions. Adding the two distance functions gives:

$$A_1^{SUM}: \quad 2dist^4 = a^4 + b^4 + c^4 + d^4$$
$$-2\left(a^2b^2 + a^2c^2 + a^2d^2 + b^2c^2 + b^2d^2 + c^2d^2\right) \quad (7.11)$$

This will not reduce to a quadratic form unless two of the variables are zero, but it will reduce to quadratic form for any two non-zero variables. Again the rotations are all space-time rotations, and so this cannot form a geometric space.

Summary of the A_1 Lie group:

We have used the A_1 Lie group to illustrate to the reader the idea that a Lie group might be just a number of intersecting circles, or intersecting hyperbolas, rather than a 'spherical

---

[34] See: Dennis Morris : Upon General Relativity

surface'. All of the $C_2 \times C_2$ Lie groups are of this 'intersecting circles' nature including the important $SU(2)$ Lie group. With a little thought, the reader will realise that any division algebra space, spinor space, with sub-algebras (derived from a finite group with sub-groups) will have a Lie group which is not a complete 'spherical surface'. There are other surprises in store, like double cover, but we have shocked the reader enough for now.

The A_2 Algebras:

The group $C_2 \times C_2$ holds six commutative algebras which have one symmetric imaginary variable and two anti-symmetric imaginary variables. These algebras are of the form:

$$\exp\left(\begin{bmatrix} a & b & c & d \\ -b & a & -d & c \\ -c & -d & a & b \\ d & -c & -b & a \end{bmatrix}\right) \qquad (7.12)$$

$$a + b\sqrt[2]{-1} + c\sqrt[2]{-1} + d\sqrt[2]{+1} \qquad (7.13)$$

The distance function, determinant of the matrix within the exponential, (7.12), of this algebra is:

$$dist^4 = \left((a-d)^2 + (b+c)^2\right)\left((a+d)^2 + (b-c)^2\right) \qquad (7.14)$$

The A_2 Lie groups are of the form:

$$PROD \left\{ \begin{bmatrix} \cos b & \sin b & 0 & 0 \\ -\sin b & \cos b & 0 & 0 \\ 0 & 0 & \cos b & \sin b \\ 0 & 0 & -\sin b & \cos b \end{bmatrix} \begin{bmatrix} \cos c & 0 & \sin c & 0 \\ 0 & \cos c & 0 & \sin c \\ -\sin c & 0 & \cos c & 0 \\ 0 & -\sin c & 0 & \cos c \end{bmatrix} \begin{bmatrix} \cosh d & 0 & 0 & \sinh d \\ 0 & \cosh d & -\sinh d & 0 \\ 0 & -\sinh d & \cosh d & 0 \\ \sinh d & 0 & 0 & \cosh d \end{bmatrix} \right\} \quad (7.15)$$

Like the A_1 Lie groups, the A_2 Lie groups are not 'spherical surfaces' but are intersecting 'circles'. Whereas the A_1 Lie groups were three intersecting hyperbolas, The A_2 Lie groups are two intersecting circles and one intersecting hyperbola. Of course, since these algebras are commutative, the Lie groups are commutative.

Double cover:
There is another property of the Lie groups within the $C_2 \times C_2$ finite group. That property is called double cover. In conventional texts on Lie groups and on spinors, double cover is defined in rather obscure ways. The obscurity derives from a lack of proper understanding of the nature of

double cover. Looking the Lie group, the rotation matrix, (7.15), and bearing in mind that in these spaces we can rotate in only one 2-dimensional plane at a time, we can set two of the three parameters to zero[35]. The Lie group then becomes of the form:

$$\begin{bmatrix} \cos b & \sin b & 0 & 0 \\ -\sin b & \cos b & 0 & 0 \\ 0 & 0 & \cos b & \sin b \\ 0 & 0 & -\sin b & \cos b \end{bmatrix} \tag{7.16}$$

Note that this is not rotation about an axis as taking the eigenvectors of the matrix will confirm.

We have two 2-dimensional Euclidean clockwise rotations within this matrix. We have twice the rotation we would have in our 4-dimensional space-time. If we feed in $2\pi's$ worth of rotation, one $\pi's$ worth into the upper left corner rotation and one $\pi's$ worth into the lower right corner rotation, we get rotation from unity to minus unity. It takes $4\pi's$ worth of rotation to rotate all the way back to unity. Not convinced? There are A_2 rotation matrices that hold sub-rotations like:

$$\begin{bmatrix} \cos b & \sin b & 0 & 0 \\ -\sin b & \cos b & 0 & 0 \\ 0 & 0 & \cos b & -\sin b \\ 0 & 0 & \sin b & \cos b \end{bmatrix} \tag{7.17}$$

---

[35] This is no more than a change of co-ordinate system.

This is a rotation in both the clockwise direction and the anti-clockwise direction at the same time. One cannot claim that these two oppositely oriented rotations are the same rotation. Each must have its own $\pi's$ worth of rotation to get to minus unity and each must have another $\pi's$ worth of rotation to get back to unity. This is the 'an electron must rotate through 720^0 to get back to unity' phenomenon.

This is commutative rotation because the A_2 algebras are commutative algebras. This is rotation in both clockwise and anti-clockwise directions at the same time.

Taking the distance function to bits:

The A_2 algebras have distance functions of the form:

$$dist^4 = \left((a-d)^2 + (b+c)^2\right)\left((a+d)^2 + (b-c)^2\right) \quad (7.18)$$

There are three 2-dimensional sub-algebras within each A_2 algebra formed from the real variable and one of the imaginary variables. The distance functions of these sub-algebras are:

$$dist^2 = a^2 + b^2$$
$$dist^2 = a^2 + c^2 \quad (7.19)$$
$$dist^2 = a^2 - d^2$$

The pairings of the imaginary variables are:

$$dist^2 = b^2 - c^2$$
$$dist^2 = b^2 + d^2 \qquad (7.20)$$
$$dist^2 = c^2 + d^2$$

We have two 2-dimensional space-time rotations and four Euclidean 2-dimensional rotations. It is not possible to form a 4-dimensional space with this mixture of 2-dimensional Euclidean rotations and 2-dimensional space-time rotations – the signs do not work. A 4-dimensional geometric space must have either:

$$dist^2 = a^2 + b^2 + c^2 + d^2$$
$$dist^2 = a^2 + b^2 + c^2 - d^2 \qquad (7.21)$$
$$dist^2 = a^2 + b^2 - c^2 - d^2$$

These correspond to:

6   Euclidean rotations

3   Euclidean rotations   &   3   space-time rotations

2   Euclidean rotations   &   4   space-time rotations

$$(7.22)$$

We cannot form a 4-dimensional geometric space from a single A_2 algebra; we have only three of the six necessary angles anyway. The $(3,3)$ case is obviously our 4-dimensional space-time.

Expectation spaces from the A_2 algebras:

The sum of the distance functions of the six A_2 algebras will not factorise to quadratic form. The distance functions of the individual A_2 algebras will not factorise to quadratic form.

There are no expectation spaces from the A_2 algebras.

Summary:

We have no great interest in the A_1 & A_2 Lie groups. We have used the A_1 & A_2 Lie groups to introduce the reader to 'intersecting circles' type of Lie groups which have less parameters, angles, than there are potential 2-dimensional planes. We have used the A_1 & A_2 Lie groups to introduce the reader to double cover. We have introduced the reader to multi-dimensional rotation as in the C_3 group of a previous chapter; this is rotation which is not about an axis. We have introduced the reader to commutative rotation in more than two dimensions.

We will be returning to the Lie groups of the $C_2 \times C_2$ group after the next chapter. When we return, we will examine the non-commutative Lie groups of the A_3 algebras and the quaternion algebras. The Lie groups of the quaternion algebras are known as $SU(2)$, and so we will meet with the conventional presentation of Lie groups. Before that, we will now look at the Lie groups of the geometric space which is our 4-dimensional space-time.

Chapter 8

Our 4-dimensional Space-time

We observe that our 4-dimensional space-time is a geometric space – we can wave our arms around in it. Our 4-dimensional space-time is a fundamentally different kind of space from the spinor spaces we find within the finite groups. It is also a much rarer type of space; it is believed that our 4-dimensional space-time is, other than the two 2-dimensional spaces, the unique geometric space. It is believed that there are no other spaces of any dimension in which we can wave our arms around, other than the 2-dimensional spaces[36]. This concurs with observation.

Geometric spaces contain geometric Lie groups, geometric continuous groups. These geometric Lie groups are fundamentally different from the spinor Lie groups. There are three geometric Lie groups within our 4-dimensional space-time; these geometric Lie groups are: $SO(2)$, $SO(3)$, and $SO(3,1)$; the latter of these is called the Lorentz group. These correspond to a circle in $\mathbb{R}^2$, a spherical surface in $\mathbb{R}^3$, and a 'semi-hyperbolic semi-spherical surface' in $\mathbb{R}^4$.

---

[36] There is a question regarding the quaternion expectation space which we will address in due course.

$SU(2)$ and $SO(3)$:

The reader will often see within the literature statements like $SU(2)$ is a simply connected double cover of $SO(3)$. This reads as if these two Lie groups were similar in some way. By the mathematical definitions of double cover and simple connection, it is true to say $SU(2)$ is a simply connected double cover of $SO(3)$, but it is extremely misleading. How misleading will become apparent when we look at $SU(2)$. The idea that these two Lie groups might be similar in some way is due to a misunderstanding of their different natures. Another common statement of the same type is the statement that there is a diffeomorphism between $SU(2)$ and $SO(3)$. Again this is true; again it is very misleading.

Lie groups in geometric space:

The spinor Lie groups of the finite groups are expressed as a single rotation matrix with $(O-1)$ parameters where O is the order of the particular finite group. Other than in 2-dimensions, we can never wave our arms around in this type of Lie group space.

We can wave our arms around in a geometric space; that is why we call it a geometric space. To be able to wave our arms around, we need to be able to rotate within the space in all possible 2-dimensional planes.

There is one 2-dimensional plane in 2-dimensional space, and so we need one parameter, angle, to be able to wave our arms around in 2-dimensional space. There are three 2-dimensional planes in 3-dimensional geometric space and so we need three parameters, angles, to be able to wave our arms around in 3-dimensional geometric space. There are six 2-dimensional planes in 4-dimensional geometric space and so we need six parameters, angles, to be able to wave our arms around in 4-dimensional geometric space.

The geometric space Lie groups are:

2-dimensional, $SO(2) \simeq U(1)$:

$$SO(2) = \begin{bmatrix} \cos\theta & \sin\theta \\ -\sin\theta & \cos\theta \end{bmatrix} \equiv \left[e^{i\theta} \right] = U(1) \qquad (8.1)$$

SO stands for special orthogonal. We see that the unitary Lie group in the spinor space of the Euclidean complex plane is the same as the special orthogonal group. There is a conceptual difference between these groups. The special orthogonal group is set in a space formed of two real variables, $\mathbb{R}^2$, whereas the unitary group is set in a spinor space of one real variable and one imaginary variable. Imaginary variables exist in only division algebras, and all division algebras, except the 1-dimensional real numbers, have imaginary variables; $\mathbb{R}^2$ is not a division algebra space. The $SO(2)$ Lie group is the one matrix (8.1).

3-dimensional, $SO(3)$, is the three matrices:

$$\begin{bmatrix} \cos\theta_1 & \sin\theta_1 & 0 \\ -\sin\theta_1 & \cos\theta_1 & 0 \\ 0 & 0 & 1 \end{bmatrix} \quad \begin{bmatrix} \cos\theta_2 & 0 & \sin\theta_2 \\ 0 & 1 & 0 \\ -\sin\theta_2 & 0 & \cos\theta_2 \end{bmatrix} \qquad (8.2)$$

and:

$$\begin{bmatrix} 1 & 0 & 0 \\ 0 & \cos\theta_3 & \sin\theta_3 \\ 0 & -\sin\theta_3 & \cos\theta_3 \end{bmatrix} \tag{8.3}$$

Note that each of these matrices has an eigenvector which is independent of the parameter, angle. This eigenvector is constant, and is the axis of rotation. This Lie group is the familiar 3-dimensional spatial part of our 4-dimensional space-time. The $SO(3)$ Lie group is the set of three matrices in (8.2) & (8.3); it is not the case that $SO(3)$ is a single matrix. Note that $SO(3)$ has three $SO(2)$ sub-groups which are 2-dimensional rotations in three different orthogonal 2-dimensional planes. We knew that from observation; we just like to see our words written in a book.

4-dimensional $SO(3,1)$ is the six matrices:

$$\begin{bmatrix} \cosh\chi_1 & \sinh\chi_1 & 0 & 0 \\ \sinh\chi_1 & \cosh\chi_1 & 0 & 0 \\ 0 & 0 & 1 & 0 \\ 0 & 0 & 0 & 1 \end{bmatrix} \quad \begin{bmatrix} \cosh\chi_2 & 0 & \sinh\chi_2 & 0 \\ 0 & 1 & 0 & 0 \\ \sinh\chi_2 & 0 & \cosh\chi_2 & 0 \\ 0 & 0 & 0 & 1 \end{bmatrix}$$

$$\tag{8.4}$$

$$\begin{bmatrix} \cosh\chi_3 & 0 & 0 & \sinh\chi_3 \\ 0 & 1 & 0 & 0 \\ 0 & 0 & 1 & 0 \\ \sinh\chi_3 & 0 & 0 & \cosh\chi_3 \end{bmatrix} \quad \begin{bmatrix} 1 & 0 & 0 & 0 \\ 0 & \cos\theta_1 & \sin\theta_1 & 0 \\ 0 & -\sin\theta_1 & \cos\theta_1 & 0 \\ 0 & 0 & 0 & 1 \end{bmatrix}$$

$$\tag{8.5}$$

$$\begin{bmatrix} 1 & 0 & 0 & 0 \\ 0 & 1 & 0 & 0 \\ 0 & 0 & \cos\theta_2 & \sin\theta_2 \\ 0 & 0 & -\sin\theta_2 & \cos\theta_2 \end{bmatrix} \quad \begin{bmatrix} 1 & 0 & 0 & 0 \\ 0 & \cos\theta_3 & 0 & \sin\theta_3 \\ 0 & 0 & 1 & 0 \\ 0 & -\sin\theta_3 & 0 & \cos\theta_3 \end{bmatrix}$$

$$(8.6)$$

There is also a $SO(4)$ which is the same as the above six $SO(3,1)$ matrices but with the hyperbolic trigonometric functions replaced by the Euclidean trigonometric functions. $SO(4)$ is six 2-dimensional Euclidean rotations in 4-dimensional space.

Note that the above set of six 2-dimensional rotations in 4-dimensional space-time, $SO(3,1)$, are all rotations about two axes. There are two constant eigenvectors of each of the six matrices. The $SO(3,1)$ Lie group is the set of six matrices in (8.4) & (8.5) & (8.6); it is not the case that $SO(3,1)$ is a single matrix. Note that $SO(3,1)$ has six 2-dimensional sub-groups; three of these are $SO(2)$ sub-groups. $SO(3,1)$ also has $SO(3)$ as a sub-group. There is also a 3-dimensional sub-group of hyperbolic rotations, changes of velocity.

It is the case that geometric spaces, and hence geometric Lie groups, always have sub-spaces, sub-Lie groups of all lesser dimensions. Geometric spaces have real variables and are of the form $\mathbb{R}^n$ rather than the form of a division algebra. Spaces of the form $\mathbb{R}^n$ are such that the nature of the space is unchanged if we 'rip off a dimension or two'. Division algebra spaces, spinor spaces, are such that we cannot randomly 'rip off a dimension or two' without destroying the algebra; only if there are sub-groups of the underlying finite group can division algebra spaces have 'a dimension or two ripped' from them.

Two types of Lie groups:

We see we have two distinct types of Lie group. Both types are continuous groups which are 'rotational surfaces' within a type of space. The two types of Lie groups are based in two types of space. One type of space is the geometric space of which the primary example is our beloved 4-dimensional space-time. The other type of space is the spinor space which is a division algebra space.

Geometric spaces have as many parameters, angles, as there are potential 2-dimensional planes, different pairs of variables, in the n-dimensional space. These are distributed over as many separate matrices as there are parameters.

Spinor spaces have as many parameters as there are imaginary variables in the algebra. These parameters are all within the single rotation matrix in the polar form of the algebra.

How many geometric spaces?:

It has been shown elsewhere[37] by your author that there are no geometric spaces of dimension greater than four[38]. The

---

[37] See: Dennis Morris : The Uniqueness of our Space-time

[38] There is an intriguing question of a possible 4-dimensional geometric sub-space of a 8-dimensional non-geometric space.

Lie groups listed above, $SO(2)$, $SO(3)$, $SO(4)$, $SO(3,1)$ are the only geometric Lie groups.

There are an infinite number of spinor Lie groups. The question facing physics is "How do the spinor Lie groups manifest themselves in our 4-dimensional space-time geometric space". The bi-directional nature of electron spin is one of these manifestations.

Summary:
We will return to examine these special orthogonal groups in later chapters. We now return to the $C_2 \times C_2$ finite group where we examine the non-commutative Lie groups.

Chapter 9

The 4-dimensional Lie Groups – Part 2

As well as the eight commutative $C_2 \times C_2$ division algebras examined above, the A_1 & A_2 algebras, there are eight non-commutative division algebras within the commutative $C_2 \times C_2$ group. Two of these are quaternion algebras, the quaternions and the anti-quaternions, and six of these are the A_3 algebras.

The quaternion Lie group:

The quaternion Lie group is $SU(2)$ although it is rarely presented in this way. The usual presentation is that, technically, the $SU(2)$ Lie algebra is isomorphic as a Lie algebra to the quaternion commutation relations. We will begin with the quaternion Lie group and derive the $SU(2)$ Lie algebra.

The quaternion rotation matrix is:

$$\begin{bmatrix} \cos\lambda & \dfrac{b}{\lambda}\sin\lambda & \dfrac{c}{\lambda}\sin\lambda & \dfrac{d}{\lambda}\sin\lambda \\[2ex] -\dfrac{b}{\lambda}\sin\lambda & \cos\lambda & -\dfrac{d}{\lambda}\sin\lambda & \dfrac{c}{\lambda}\sin\lambda \\[2ex] -\dfrac{c}{\lambda}\sin\lambda & \dfrac{d}{\lambda}\sin\lambda & \cos\lambda & -\dfrac{b}{\lambda}\sin\lambda \\[2ex] -\dfrac{d}{\lambda}\sin\lambda & -\dfrac{c}{\lambda}\sin\lambda & \dfrac{b}{\lambda}\sin\lambda & \cos\lambda \end{bmatrix} \quad (9.1)$$

$$\lambda = \sqrt{b^2 + c^2 + d^2}$$

This is a spinor rotation. The quaternions have three 2-dimensional sub-algebras because the $C_2 \times C_2$ group has three order two sub-groups. We have only three parameters, angles, in this 4-dimensional space, and so we can rotate in only three 2-dimensional planes. The quaternion Lie group is three intersecting circles.

Looking at (9.1), we see that we have double cover rotation in each variable.

$SU(2)$ and $SO(3)$ compared:

We reiterate that the quaternion Lie group, $SU(2)$, (9.1), is just three intersecting circles. This Lie group is very different from the familiar spherical surface which is $SO(3)$ in $\mathbb{R}^3$, (8.2), above. The reader now sees that, even if it is true, how misleading it is to say that $SU(2)$ is a simply connected double cover of $SO(3)$. The two groups both have three angle parameters; the types of rotation are Euclidean in both cases. Other than the double cover, we have three 2-dimensional Euclidean rotations in both groups, and this is the basis of the diffeomorphism between these Lie groups. The quaternions are simply connected because they are a division algebra and the rotation is done with one rotation matrix whereas the $SO(3)$ rotation is not simply connected because it is done by three separate rotation matrices.

Of course, quaternion space differs from $SO(3)$ space by being 4-dimensional rather than 3-dimensional. These two Lie algebras are very different from each other.

To continue:

A repeated word of warning: The quaternion Lie group looks like a set of intersecting circles to we who sit in our geometric space. If we were sitting in quaternion space, The Lie group would appear, I presume, as a continuous rotational surface.

Because we can rotate in only one of the three 2-dimensional planes at a time, we can fool ourselves into thinking it makes sense to split the quaternion rotation matrix into three bits by setting two of the variables to zero. For example, setting $c = d = 0$ gives:

$$\begin{bmatrix} \cos b & \sin b & 0 & 0 \\ -\sin b & \cos b & 0 & 0 \\ 0 & 0 & \cos b & -\sin b \\ 0 & 0 & \sin b & \cos b \end{bmatrix} \quad (9.2)$$

This matrix will be generated by taking the exponential of the matrix:

$$U_b = \begin{bmatrix} 0 & 1 & 0 & 0 \\ -1 & 0 & 0 & 0 \\ 0 & 0 & 0 & -1 \\ 0 & 0 & 1 & 0 \end{bmatrix} \tag{9.3}$$

The other two separate variables give:

$$U_c = \begin{bmatrix} 0 & 0 & 1 & 0 \\ 0 & 0 & 0 & 1 \\ -1 & 0 & 0 & 0 \\ 0 & -1 & 0 & 0 \end{bmatrix} \quad \& \quad U_d = \begin{bmatrix} 0 & 0 & 0 & 1 \\ 0 & 0 & -1 & 0 \\ 0 & 1 & 0 & 0 \\ -1 & 0 & 0 & 0 \end{bmatrix} \tag{9.4}$$

We might then claim that these three matrices presented above, (9.3) & (9.4), are the generator matrices of the three independent 2-dimensional rotations in quaternion space. However, there are two problems with separating the generator matrices. Firstly, the quaternion rotations are not independent 2-dimensional rotations; rotation in quaternion space is 4-dimensional. Secondly, the quaternion variables are not commutative. We cannot take the exponential of the separate generators and simply multiply them together to get the quaternion rotation. More properly, we should recognise that the quaternion is a spinor space with a single rotation matrix and present the generator as:

$$\begin{bmatrix} 0 & 1_b & 1_c & 1_d \\ -1_b & 0 & -1_d & 1_c \\ -1_c & 1_d & 0 & -1_b \\ -1_d & -1_c & 1_b & 0 \end{bmatrix} \quad (9.5)$$

However, let us continue on our foolish way. Using the equivalence:

$$a + ib \equiv \begin{bmatrix} a & b \\ -b & a \end{bmatrix} \quad (9.6)$$

and the block multiplication property of matrices, allows us to reduce the proposed generator matrices, (9.3) & (9.4), to 2×2 matrices giving:

$$u_b = \begin{bmatrix} i & 0 \\ 0 & -i \end{bmatrix}, \quad u_c = \begin{bmatrix} 0 & 1 \\ -1 & 0 \end{bmatrix}, \quad u_d = \begin{bmatrix} 0 & i \\ i & 0 \end{bmatrix} \quad (9.7)$$

These matrices have imaginary eigenvalues (they all have the eigenvalues $\{i, -i\}$. These are the generator matrices. In general, for a unitary representation of the Lie group, the generator matrices are anti-hermitian (anti-symmetric). We make them hermitian (symmetric) by multiplying by i.

Multiplying by $-i$ gives:

$$u_b^* = \begin{bmatrix} 1 & 0 \\ 0 & -1 \end{bmatrix}, \quad u_c^* = \begin{bmatrix} 0 & -i \\ i & 0 \end{bmatrix}, \quad u_d^* = \begin{bmatrix} 0 & 1 \\ 1 & 0 \end{bmatrix} \quad (9.8)$$

These matrices now have real eigenvalues which are observables. These are the Pauli matrices which are conventionally called the spin $\dfrac{1}{2}$ representation of the Lie

algebra $SU(2)$. We looked at the finite group $C_2 \times C_2$; we took the Lie groups of that finite group; we separated the quaternion Lie group into its separate parts, separate circles, forgot about double cover, forgot about the non-commutative nature of the quaternions, multiplied by i, and out popped the Lie algebra $SU(2)$. Of course the commutation relations of the quaternion are the same as the commutation relations of $SU(2)$; this is established and accepted understanding. Time for some poetry:

> Little Jack Horner
>
> Sat in a corner eating a C_2 cross C_2 pie
>
> At his first chew
>
> He found $SU(2)$
>
> And said, "What a good boy am I"

Reiteration:

Having enjoyed such excellent poetry, we should gather out thoughts.

A Lie group is a 'rotational surface'. Within the spinor algebras, a Lie group is intersecting circles (could be a hyperbola or two as well). With no regard for the niceties of quaternion algebraic structure, separating each of the 'circles' leads to separate rotations which correspond to separate generators of those rotations (through the exponential). The generators of the separate rotations are the elements of the Lie algebra conventionally associated with the given Lie group.

The reader might now be thinking "Hm! we had separate matrices for each 2-dimensional rotation in the geometric space that is our 4-dimensional space-time". We can see that, if all of humanity believes that 2-dimensional rotations are the only type of rotations and that all spaces are geometric spaces, then it is understandable that all humanity could presume that a space with three 2-dimensional rotations could be represented as three generators of those rotations.

Chapter 10

The Commutator

Ought not something that bears the appellation 'algebra' to have some kind of multiplication operation? Yes – most perspicacious of the reader to see this.

We are familiar with how to multiply matrices together within commutative algebras. For more than a century, it has been assumed that multiplication within a non-commutative algebra ought to be identical to multiplication within a commutative algebra, but there are good reasons to think a different form of multiplication might be appropriate within a non-commutative algebra. Instead of the product of two non-commutative matrices, $A \& B$, being simply AB, perhaps the natural product of two matrices within a non-commutative algebra should be both:

$$
\begin{aligned}
PROD\{A, B\} &= \{AB + BA\} \\
PROD[A, B] &= [AB - BA]
\end{aligned}
\qquad (10.1)
$$

The topmost of these is called the anti-commutator. The bottommost of these is called the commutator.

Lie algebras deal with non-commutative linear transformations (that's posh talk for non-commutative matrices), and the multiplication operation of a Lie algebra is the commutator. Consider the matrices, taken from (9.8):

$$p_b = \frac{1}{2}\begin{bmatrix} 1 & 0 \\ 0 & -1 \end{bmatrix}, \quad p_c = \frac{1}{2}\begin{bmatrix} 0 & -i \\ i & 0 \end{bmatrix}, \quad p_d = \frac{1}{2}\begin{bmatrix} 0 & 1 \\ 1 & 0 \end{bmatrix} \quad (10.2)$$

We have relations like:

$$[p_b, p_c] = p_b p_c - p_c p_b = -i p_d \quad (10.3)$$

Other than having to multiply by $-i$, the set of matrices is closed under commutator multiplication. Better still, using half of the original matrices (9.7) gives:

$$[u_b, u_c] = u_d, \qquad [u_d, u_b] = u_c, \qquad [u_c, u_d] = u_b \quad (10.4)$$

We have no need of the messy $-i$; of course we don't; these are the imaginary units of the quaternions usually written as $\{\hat{i}, \hat{j}, \hat{k}\}$. The messy $-i$ arises for historical reasons because the conventional approach to Lie groups was through Lie algebras which is from the opposite direction to that from which we have approached Lie groups in this book. Along with the commutator equations, (10.4), we have the clean anti-commutator result:

$$\{u_b, u_c\} = 0, \qquad \{u_d, u_b\} = 0, \qquad \{u_c, u_d\} = 0 \quad (10.5)$$

What is a Lie algebra?:

A Lie algebra is a set of square matrices which are such that:

1) when these matrices are multiplied by a parameter (angle), say θ, the exponential of each is a 2-dimensional rotation matrix. These matrices will necessarily have zero trace because rotation

matrices (of any dimension) have a determinant of unity[39].

2) These matrices are connected to each other by the commutator in such a way that the commutator of any two of the matrices gives another of the matrices (multiplied by a number which might be complex). The particular set of commutation relations is often called the Lie algebra, and any two different sets of matrices[40] which have the same commutation relations are said to be isomorphic as Lie algebras.

It is important to realise that the commutator is such that the commutator of two of the matrices is a sum of two copies of a single other matrix and not a sum of two different matrices. When we look at the finite group S_3, we will find a commutator which is the sum of two different other matrices. We will find he same in other finite groups.

An important point:

We will see later that the requirement that the commutator of two matrices is the sum of two copies of the same other matrix is really the requirement that the matrices are within a 4-dimensional non-commutative spinor algebra of the $C_2 \times C_2$ finite group. There are only two different types of such algebras, the quaternions and the A_3 algebras. An

---

[39] The determinant of the exponential of a matrix with zero trace is unity.

[40] We do not count the same matrices written in different bases to be different Lie algebras.

implication of this is that the conventional Lie group $SU(3)$ is really an intermixing of several of these 4-dimensional quaternion Lie groups and A_3 Lie groups.

Quaternion oddments:

A quaternion can be written as a pair of Euclidean complex numbers. This notation is obfuscating, but it is much used in particle physics. We often see texts presenting the Lie algebra $SU(2)$ as[41]:

$$SU(2) = \begin{bmatrix} \mathbb{C}_1 & \mathbb{C}_2 \\ -\mathbb{C}_2^* & \mathbb{C}_1^* \end{bmatrix} \quad : \quad \mathbb{C}_1\mathbb{C}_1^* + \mathbb{C}_2\mathbb{C}_2^* = 1 \quad (10.6)$$

Using the block multiplication property of matrices, we see that this is:

$$\begin{bmatrix} a & b & c & d \\ -b & a & -d & c \\ -c & d & a & -b \\ -d & -c & b & a \end{bmatrix} \quad : \quad a^2 + b^2 + c^2 + d^2 = 1$$

$$(10.7)$$

It is a quaternion of unit length. The set of quaternions of unit length is the quaternion rotation matrix, (9.1).

Particle physicist often use Weyl spinors. Particle physicists are accustomed to thinking of spinors as a pair (or two pairs)

---

[41] Eric Weisstein: CRC Concise Encyclopedia of Mathematics

of complex numbers. A Weyl spinor is just two complex numbers presented as a column vector:

$$\begin{bmatrix} a + ib \\ c + id \end{bmatrix} \tag{10.8}$$

Associated with this obscure notation is a strange way of calculating the inner product and norm of this Weyl spinor. The upshot is that the Weyl spinor is just a quaternion in strange notation with a strange inner product. It would be more appropriate to write the Weyl spinor as:

$$\begin{bmatrix} a + ib \\ jc + kd \end{bmatrix} \tag{10.9}$$

If the components of the Weyl spinor were just two complex numbers, they would commute. It is an often ignored property of Weyl spinors that, within a Weyl spinor, the two components do not commute, and so (10.8) is deceptive.

Weyl spinors are taken by particle physicists to represent electrons[42]. Quaternions have double cover; they have a Lie group of intersecting circles corresponding to the quantum directional nature of electron spin, and they have the $SU(2)$ commutation relations. It seems that except for the obfuscating notation, particle physicists have it right.

Note: Because particle physicists write a quaternion as a pair of complex numbers in a vector, they have convinced themselves that there exist vector spaces with complex axes. Upon these complex vector spaces, the physicists put an

---

[42] Fermions in general, actually, but electrons sounds simpler.

inner product. Such spaces are called complex Hilbert spaces. Complex Hilbert spaces are a central concept in modern quantum physics. In the view of the new approach mathematics, complex Hilbert spaces do not exist. However, all is not lost for quaternions do exist.

Chapter 11

The 4-dimensional Lie Groups – Part 3

The anti-quaternion Lie group:

Many ages past, lost in antiquity now, it was mentioned to the erudite and pains-taking reader that there are two quaternion algebras within the finite group $C_2 \times C_2$. The anti-quaternions are algebraically isomorphic to the quaternions but the $SU(2)$ commutation relations are reversed. Within the quaternion algebra, we have, in conventional notation:

$$\hat{i}\hat{j} = \hat{k}, \qquad \hat{j}\hat{k} = \hat{i}, \qquad \hat{k}\hat{i} = \hat{j}$$
$$\hat{j}\hat{i} = -\hat{k}, \qquad \hat{k}\hat{j} = -\hat{i}, \qquad \hat{i}\hat{k} = -\hat{j}$$

(11.1)

Within the anti-quaternion algebra, we have:

$$\hat{i}\hat{j} = -\hat{k}, \qquad \hat{j}\hat{k} = -\hat{i}, \qquad \hat{k}\hat{i} = -\hat{j}$$
$$\hat{j}\hat{i} = \hat{k}, \qquad \hat{k}\hat{j} = \hat{i}, \qquad \hat{i}\hat{k} = \hat{j}$$

(11.2)

The commutations are reversed. If a quaternion, Weyl spinor, represents an electron, it seems that we must have both right-chiral electrons and left-chiral electrons. There is no mass in quaternion space, and so our electrons are massless. This fits with having both left-chiral and right-chiral electrons. It seems that the massive electron will emerge when we superimpose the quaternions and the anti-quaternions – take the expectation space, but this is not yet properly understood. Of course, the concept of right-chiral electrons and left-chiral electrons underpins the electro-weak unification part of particle physics.

The anti-quaternion Lie group is:

$$\begin{bmatrix} \cos\lambda & \dfrac{b}{\lambda}\sin\lambda & \dfrac{c}{\lambda}\sin\lambda & \dfrac{d}{\lambda}\sin\lambda \\[2mm] -\dfrac{b}{\lambda}\sin\lambda & \cos\lambda & \dfrac{d}{\lambda}\sin\lambda & -\dfrac{c}{\lambda}\sin\lambda \\[2mm] -\dfrac{c}{\lambda}\sin\lambda & -\dfrac{d}{\lambda}\sin\lambda & \cos\lambda & \dfrac{b}{\lambda}\sin\lambda \\[2mm] -\dfrac{d}{\lambda}\sin\lambda & \dfrac{c}{\lambda}\sin\lambda & -\dfrac{b}{\lambda}\sin\lambda & \cos\lambda \end{bmatrix} \quad (11.3)$$

$$\lambda = \sqrt{b^2 + c^2 + d^2}$$

This is different from the quaternion Lie group only in the distribution of minus signs. However that distribution of minus signs hides a little secret. The quaternions are not a perfect reflection of the anti-quaternions. Most remarkably, within the perfectly symmetrical $C_2 \times C_2$ group, we have a little imbalance in the nature of the double cover.

Double cover comes in two forms. One form is rotation in both clockwise and anti-clockwise directions at the same time; the other form is rotation twice in the same direction.

Looking at the individual variables in the anti-quaternion rotation matrix, (11.3), we see the b variable rotates twice in the clockwise direction, the d variable also rotates twice in the clockwise direction, and the c variable rotates in both clockwise and anti-clockwise directions.

Looking at the individual variables in the quaternion rotation matrix, (9.1), we see the b variable rotates in both the clockwise direction and the anti-clockwise direction, the d variable also rotates in both the clockwise direction and the anti-clockwise direction, and the c variable rotates twice in the clockwise direction.

If we associate rotation with charge, as is done in quantum field theory[43], we might say that the variables which rotate in both directions at once have no charge. This would mean that the anti-quaternion electron has twice as much charge (not electric charge) as the quaternion electron – there's a thing. There's a thing because, in quantum field theory, a right-chiral spinor such as a right-chiral electron has twice as much weak hypercharge, Y, as a left-chiral spinor such as a left-chiral electron. We must point out that a right-chiral electron has isospin charge $I^3 = 0$ whereas a left-chiral electron has isospin charge $I^3 = -\frac{1}{2}$. Looking at the quaternions and anti-quaternions, the nature of isospin charge is not obvious, but we do have the electroweak Gell-Mann-Nishijima relation:

$$Q = I^3 + \frac{Y}{2} \tag{11.4}$$

Wherein Q is the electric charge of the electron.

We add for completeness that a quaternion and an anti-quaternion commute with each other. The weak isospin

---

[43] In quantum field theory, each generator of a Lie algebra is associated with some kind of conserved charge.

charge and the weak hypercharge also commute with each other.

There is another difference between the quaternions and the anti-quaternions. The quaternion algebra matrix form is four 2×2 blocks which are complex numbers. There are only two such complex number blocks within the anti-quaternion algebra matrix form. The other two anti-quaternion blocks are complex numbers in a different basis, but it seems that basis is important. Perhaps only 'complex number' blocks can be manifest in our 4-dimensional space-time. Thus, we might expect twice as many fermions from the quaternion algebra as from the anti-quaternion algebra. The absence of right-handed neutrinos comes to mind. We are now getting very speculative.

The matter of weak hypercharge and isospin is not properly understood within the quaternion algebras. The interesting bit is that within an entirely symmetrical finite group, we have an imbalance. Makes one think of the left-handed nature of the weak force, does it not.

The emergent quaternion expectation space:
The quaternions and the anti-quaternions have the distance functions:

$$\begin{aligned}
dist_Q^4 &= \left(a^2 + b^2 + c^2 + d^2\right)^2 \\
dist_{AQ}^4 &= \left(a^2 + b^2 + c^2 + d^2\right)^2
\end{aligned} \tag{11.5}$$

These obviously reduce to quadratic form. Adding these forms the quaternion expectation distance function:

$$dist^2 = a^2 + b^2 + c^2 + d^2 \qquad (11.6)$$

This distance function allows 2-dimensional Euclidean rotation between all six pairs of variables. Given enough angle parameters, we could wave our arms in this expectation space. It is however, a timeless space. The signs before the squared variables are all pluses, and so we have no time in this space. In such a space, we might have instantaneous collapse of a spatially extensive wave-function of instantaneous communication between spatially separated entities.

The quaternion emergent expectation space has the same distance function as the quaternion space, and so the distance function alone is not sufficient to guarantee that the emergent expectation space is not a division algebra.

It is not understood how the quaternion emergent expectation space fits into the physical universe. Of course, we need six angle parameters to form a geometric space. We have two sets of three angles, but perhaps there is more to 'fitting together' two quaternions than just the number of angle parameters.

Chapter 12

The 4-dimensional Lie Groups – Part 4

The A_3 Lie groups:

The six A_3 algebras are important because they hold as an emergent expectation space the geometric space that is our 4-dimensional space-time. Of the six A_3 algebras, there are three pairs comprised, like the quaternion algebras, of an algebra and its anti-algebra. Again there is the double cover imbalance between an algebra and its anti-algebra. There is also a 'hyperbolic complex number block' imbalance between an algebra and its anti-algebra.

First we will look at the Lie groups, continuous groups, associated with these algebras. A typical Lie A_3 group is:

$$SAS_{Rot} = \exp\left(\begin{bmatrix} 0 & b & c & d \\ b & 0 & d & c \\ -c & d & 0 & -b \\ d & -c & -b & 0 \end{bmatrix}\right) \tag{12.1}$$

This is:

$$
\begin{bmatrix}
\cosh \lambda & \dfrac{b}{\lambda}\sinh \lambda & \dfrac{c}{\lambda}\sinh \lambda & \dfrac{d}{\lambda}\sinh \lambda \\[2.5ex]
\dfrac{b}{\lambda}\sinh \lambda & \cosh \lambda & \dfrac{d}{\lambda}\sinh \lambda & \dfrac{c}{\lambda}\sinh \lambda \\[2.5ex]
-\dfrac{c}{\lambda}\sinh \lambda & \dfrac{d}{\lambda}\sinh \lambda & \cosh \lambda & -\dfrac{b}{\lambda}\sinh \lambda \\[2.5ex]
\dfrac{d}{\lambda}\sinh \lambda & -\dfrac{c}{\lambda}\sinh \lambda & -\dfrac{b}{\lambda}\sinh \lambda & \cosh \lambda
\end{bmatrix}
$$

$$\lambda = \sqrt{b^2 - c^2 + d^2}$$

$$(12.2)$$

Although it is not immediately apparent, this Lie group has a (double cover) Euclidean rotation within it; setting $b = d = 0$, gives:

$$
\begin{bmatrix}
\cos c & 0 & \sin c & 0 \\
0 & \cos c & 0 & \sin c \\
-\sin c & 0 & \cos c & 0 \\
0 & -\sin c & 0 & \cos c
\end{bmatrix}
\qquad (12.3)
$$

Each of the other variables corresponds to a (double cover) hyperbolic rotation. Because the A_3 algebra is an algebra, it mixes these two types of rotation leading to the product of two hyperbolic rotations being a Euclidean rotation:

$$\begin{bmatrix} 0 & b & 0 & 0 \\ b & 0 & 0 & 0 \\ 0 & 0 & 0 & -b \\ 0 & 0 & -b & 0 \end{bmatrix} \begin{bmatrix} 0 & 0 & 0 & d \\ 0 & 0 & d & 0 \\ 0 & d & 0 & 0 \\ d & 0 & 0 & 0 \end{bmatrix}$$

$$= \begin{bmatrix} 0 & 0 & bd & 0 \\ 0 & 0 & 0 & bd \\ -bd & 0 & 0 & 0 \\ 0 & -bd & 0 & 0 \end{bmatrix}$$

(12.4)

Of course, the reader familiar with the Lorentz group, $SO(3,1)$, will recall that this property of making a Euclidean rotation from two space-time rotations (boosts) is a much applauded property of the Lorentz group.

Again we have only three parameters in a 4-dimensional space, and so the Lie group, as seen from our 4-dimensional space-time, is an intersecting circle with two intersecting hyperbolas. Seen from its own space, the Lie group will appear to be a continuous surface, or so your author presumes without any observational justification for doing so.

This single A_3 algebra has three variables which might be viewed as generators. Of course, since this is a spinor space, these variables, generators, ought not to be separated but ought to be a single matrix; non-the-less, we continue and we separate the variables, generators, as we did in the quaternion case above.

$$\begin{bmatrix} 0 & 1 & 0 & 0 \\ 1 & 0 & 0 & 0 \\ 0 & 0 & 0 & -1 \\ 0 & 0 & -1 & 0 \end{bmatrix}, \begin{bmatrix} 0 & 0 & 1 & 0 \\ 0 & 0 & 0 & 1 \\ -1 & 0 & 0 & 0 \\ 0 & -1 & 0 & 0 \end{bmatrix}, \begin{bmatrix} 0 & 0 & 0 & 1 \\ 0 & 0 & 1 & 0 \\ 0 & 1 & 0 & 0 \\ 1 & 0 & 0 & 0 \end{bmatrix}$$

$$(12.5)$$

The Lorentz group algebra $SO(3,1)$ has six generators. Clearly, this single A_3 algebra cannot be the Lorentz group, but it is part of the Lorentz group.

Above, (8.4) & (8.5) & (8.6), we gave the six 2-dimensional rotations in our 4-dimensional space-time and said that it was the Lie group $SO(3,1)$. That is true, but it is not the Lie algebra of $SO(3,1)$. The algebra of $SO(3,1)$ is the generators of the six 2-dimensional rotations in our 4-dimensional space-time together with the commutation relations of those generators.

The six generators of the Lorentz Lie algebra $SO(3,1)$ are conventionally given as:

$$J_1 = \begin{bmatrix} 0 & 0 & 0 & 0 \\ 0 & 0 & 0 & 0 \\ 0 & 0 & 0 & -i \\ 0 & 0 & i & 0 \end{bmatrix} \qquad J_2 = \begin{bmatrix} 0 & 0 & 0 & 0 \\ 0 & 0 & 0 & -i \\ 0 & 0 & 0 & 0 \\ 0 & i & 0 & 0 \end{bmatrix} \quad (12.6)$$

$$J_3 = \begin{bmatrix} 0 & 0 & 0 & 0 \\ 0 & 0 & i & 0 \\ 0 & -i & 0 & 0 \\ 0 & 0 & 0 & 0 \end{bmatrix} \qquad K_1 = \begin{bmatrix} 0 & 1 & 0 & 0 \\ 1 & 0 & 0 & 0 \\ 0 & 0 & 0 & 0 \\ 0 & 0 & 0 & 0 \end{bmatrix} \qquad (12.7)$$

$$K_2 = \begin{bmatrix} 0 & 0 & 1 & 0 \\ 0 & 0 & 0 & 0 \\ 1 & 0 & 0 & 0 \\ 0 & 0 & 0 & 0 \end{bmatrix} \qquad K_3 = \begin{bmatrix} 0 & 0 & 0 & 1 \\ 0 & 0 & 0 & 0 \\ 0 & 0 & 0 & 0 \\ 1 & 0 & 0 & 0 \end{bmatrix} \qquad (12.8)$$

The K_i generators of the Lorentz group are generators of 2-dimensional space-time rotations, hyperbolic rotations, in 4-dimensional space-time – just take the exponential. The J_i generators of the Lorentz group are generators of 2-dimensional Euclidean rotations in 4-dimensional space-time that have been multiplied by $i = \sqrt[2]{-1}$ – divide by i and take the exponential.

We see that two of the A_3 generators above, (12.5), are double covers of K_1 & K_3. It is not so obvious that the third A_3 generator given above is a double cover of J_2 because there is a change, accomplished by multiplication by $-i$, involved and the essential geometric essence is placed differently within the matrix. None-the-less, the other A_3 generator is a double cover of J_2.

The Lorentz group is the Lie group of a geometric space. The Lie groups of the A_3 algebras are Lie groups of a spinor space. The generators of the spinor space are generators of double cover rotations. The generators of the Lorentz group are generators of single cover rotations.

Although we have shown only that one A_3 algebra produces generators corresponding, other than by double cover, to some of the Lorentz group generators, all A_3 algebras produce similar double cover generators of the Lorentz group generators. Taken together, the A_3 algebras produce double cover versions of all the Lorentz group generators. They also produce a double cover anti-Lorentz group analogously to the anti-quaternions. There are commutation relations between the Lorentz group generators; these are matched by the set of A_3 generators.[44]

Above, we saw the generators of the $SU(2)$ Lie algebra fall directly out of the quaternion space. The Lorentz group Lie algebra generators fall out of the set of A_3 algebras in a more complicated way, but, none-the-less, we have derived the Lorentz group Lie algebra from the $C_2 \times C_2$ finite group.

Expectation space of the A_3 algebras:

The distance functions of the six A_3 algebras sum to produce the expectation distance function of these algebras. An individual A_3 algebras has a distance function of the form:

---

[44] The details of the commutation relations can be found in Dennis Morris : The Physics of Empty Space.

$$dist^4 = \left(t^2 + x^2 - y^2 - z^2\right)^2$$
$$dist^2 = t^2 + x^2 - y^2 - z^2 \tag{12.9}$$

We have:

$$Sum \begin{cases} t^2 + x^2 - y^2 - z^2 \\ t^2 + x^2 - y^2 - z^2 \\ t^2 - x^2 - y^2 + z^2 \\ t^2 - x^2 - y^2 + z^2 \\ t^2 - x^2 + y^2 - z^2 \\ t^2 - x^2 + y^2 - z^2 \end{cases} \rightarrow \tag{12.10}$$

$$6\ dist^2 = 6t^2 - 2x^2 - 2y^2 - 2z^2$$

The 6 is just a scaling factor, and the 3 is just the units in which we measure time or space. We have the distance function of our 4-dimensional space-time.

$$dist^2 = t^2 - x^2 - y^2 - z^2 \tag{12.11}$$

This space allows 2-dimensional rotation in all possible 2-dimensional planes – we can wave our arms around in this space. The Lie group of this space is given above, (8.4) & (8.5) & (8.6), as $SO(3,1)$.

Unlike the quaternion expectation geometric space, this distance function is not the distance function of a division algebra – the form of it is not closed under multiplication and so it cannot be the determinant of a matrix which is closed under multiplication.

Your author has shown elsewhere[45] how the affine connection, curvature, of our 4-dimensional space-time and the metric tensor of our 4-dimensional space-time emerge from the A_3 algebras. Such matters are not the concern of this book, but let us consider one question.

Why is it that we observe our 4-dimensional space-time around us but we do not observe the quaternion expectation space around us? We do not have a clear understanding of this, however, we opine that it might be something to do with the number of algebras. Although a quaternion angle is defined by three real numbers, within a particular algebra, these three numbers are tied together to form the angle. Since we have only two quaternion algebras, we effectively have only two angles, even if they are quaternion angles. Since there are six A_3 algebras, we have six angles. We need six angles to form a 4-dimensional geometric space. The reader will form their own opinion.

Summary:

Other than $SU(3)$, we have seen the Lie algebras familiar to physicists emerge from the finite groups. These algebras fell from the continuous rotation groups as the generators of the single 2-dimensional rotations. We have not dwelt overly upon the commutators of these generators; that is dealt with in many other texts or can be simply calculated by the reader.

---

[45] See: Dennis Morris : Upon General Relativity

As we look at higher dimensions, we might expect a similar pattern of each finite group having Lie groups presented as rotation matrices. This certainly is the case, but there is much unexplored territory here.

As we look at higher dimensions, we might expect a similar pattern of geometric spaces falling out of the finite groups as emergent expectation spaces with their own particular geometric Lie groups. This does not happen. Your author has shown elsewhere[46] that there are no finite groups of order greater than four that hold geometric spaces. Thus, all the geometric Lie groups are presented above.

Preview of the rest of this book:

Immediately, the reader cries, "Where does $SU(3)$ fit into this? We need $SU(3)$ to mix the quark colours in QCD". It is a good question, and it is a question which we are unable to answer fully. There is much in the conventional presentation of Lie group theory which we have barely touched upon – Dynkin diagrams etc. How do we tie our, now clarified, understanding of Lie groups into the conventional mathematics associated with Lie groups? Not easily and perhaps not satisfactorily. Later in this book we will look at the conventional treatment of Lie groups and Lie algebra from the perspective we have reached through understanding Lie groups. In preparation for that, we will first look at higher dimensional Lie groups.

---

[46] See: Dennis Morris : The Uniqueness of our Space-time

Chapter 13

Lie Groups of Higher Dimensions

There are spinor spaces, division algebras, types of complex number - all the same things - of every dimension because there are spinor spaces in every finite group. The distribution of the numbers and types of these spinor spaces is not known nor completely understood. There are some surprises; for example the order eight dicyclic group (also known as the quaternion group) has within it 128 copies of only one type of spinor space comprised of a real number and one square root of plus unity and six fourth roots of plus unity. The order fifteen cyclic group contains 125 different spinor spaces. Each of these spinor spaces, division algebras, has a polar form and hence has a rotation matrix which is a continuous group, a spinor Lie group.

There are no expectation spaces of dimension greater than four which are able to support 2-dimensional rotation in every 2-dimensional plane; in other words, there are no geometric spaces of dimension greater than four. An astute observer of the physical universe might already have noticed there are no five or six or seven or … dimensional spaces in our universe.

There is however a peculiarity in the order eight $C_2 \times C_2 \times C_2$ group in that it holds one spinor space such that, if the eight variables have values in pairs such that $\{e = a, f = b, g = c, h = d\}$, then the 8-dimensional expectation space becomes a 4-dimensional geometric space; it seems as if the 8-dimensional space is folding into a 4-dimensional geometric space. This phenomenon is poorly understood, and its implications are unconsidered. With this in the back of our mind, we will look at the Lie groups of the finite group $C_2 \times C_2 \times C_2$. For completeness, we will also consider Lie groups of dimensions five, six and seven and the other order eight finite groups. There is a great insight within the order six groups and the other order eight groups regarding the nature of the commutation relations of a non-commutative spinor space.

We also have in the back of our mind that the strong force of quantum field theory is associated with the conventional Lie group $SU(3)$, and we have found no sign of this conventional Lie group within the spinor spaces of dimension four or less.

5-dim and 7-dim Lie groups:
There is only one order five finite group; this is the cyclic group C_5. In general, for any prime number, there is one and only one finite group of that order and that group is a cyclic group.

There are five non-isomorphic spinor spaces of dimension five within the finite group C_5. The five dimensional spinor spaces are formed from one real number and all possible combinations of fifth roots of plus unity or minus unity:

$$1 + 4\sqrt[5]{+1}$$
$$1 + 3\sqrt[5]{+1} + \sqrt[5]{-1}$$
$$1 + 2\sqrt[5]{+1} + 2\sqrt[5]{-1} \qquad\qquad (13.1)$$
$$1 + \sqrt[5]{+1} + 3\sqrt[5]{-1}$$
$$1 + 4\sqrt[5]{-1}$$

The order seven cyclic group holds an analogous set of 7-dimensional spinor spaces based on the seventh roots of plus or minus unity.

All cyclic groups of prime order have no sub-groups, and so rotation in these spaces is simply rotation of a nature equal to the order of the finite group analogously to the case of the 3-dimensional rotation presented above when we considered the finite group C_3.

The cyclic groups are all commutative, and the division algebras, spinor algebras, and thus the rotations in these spaces are all commutative.

Lie groups in 6-dimensional spaces:
There are two order six finite groups. One of these is the commutative group cyclic group C_6 which holds six non-isomorphic spinor spaces. There is nothing about the C_6 Lie groups which we have not previously encountered, and so we shall not dwell longer within the C_6 realm.

The other order six group is the non-commutative symmetric group S_3.

There are six non-isomorphic spinor spaces within the S_3 group. These are each comprised of a real number, two cube roots of plus unity or of minus unity, and three square roots of plus unity or of minus unity. Being a non-commutative group means that there are commutation relations associated with S_3. Now comes the great insight. The commutation relations of S_3 are quite different to the commutation relations of conventional Lie algebras. We have:

$$S_3 = \begin{bmatrix} A & B & C & D & E & F \\ C & A & B & E & F & D \\ B & C & A & F & D & E \\ D & E & F & A & B & C \\ E & F & D & C & A & B \\ F & D & E & B & C & A \end{bmatrix} \tag{13.2}$$

The commutation relations are:

$$[A,B]=0, \quad [A,C]=0, \quad [A,\mathrm{D}]=0$$
$$[A,E]=0, \quad [A,F]=0, \quad [B,C]=0 \tag{13.3}$$

And:

$$[B,D]=F-E=-[C,D]$$
$$[B,E]=D-F=-[C,E] \tag{13.4}$$
$$[B,F]=E-D=-[C,F]$$

And:

$$[D, E] = B - C = [E, F] = -[D, F] \qquad (13.5)$$

This is a quite different kind of commutation algebra (Lie algebra) from that to which we are accustomed. Within the quaternions, we find the commutator of two variables is equal to only one of the other variables. We see the commutator within the S_3 algebras is equal to a difference of two other variables. This is not difficult to understand.

Within the quaternions, we have three imaginary variables, $\left\{ \hat{i}, \hat{j}, \hat{k} \right\}$. However we multiply two of these together, we get the third of these for there are no other imaginary variables that the product could equal. This is just multiplicative closure of the elements of the underlying finite group. Within the S_3 group, we have five imaginary variables. When we take the product of any two of these, there are three other imaginary variables from which to choose[47]. Hardly surprising that we get the commutator equal to the sum of two different imaginary variables. It would be quite surprising if, when we have more than three imaginary variables, we did get the commutator equal to just one imaginary variable as happens in the quaternions.

The reader might expect that we will find this 'previously unseen' type of commutation relations in every non-commutative spinor space of dimension more than four. What would it take for a higher order spinor space to have commutation relations like the quaternions. It would take every pair of imaginary variables to be in a sub-group with

---

[47] Hm! choose is not really the right word.

the identity and one other imaginary variable – an order four sub-group. But both order four finite groups are commutative – they do not have commutation relations. Aha! The $C_2 \times C_2$ group contains non-commutative spinor spaces. We need every imaginary variable to be an element of a $C_2 \times C_2$ sub-group of the finite group which underlies the spinor space. More than that, we need every imaginary variable to be an element of a non-commutative $C_2 \times C_2$ sub-algebra of the division algebra which is the spinor space. Does this ever happen?

The 8-dimensional $C_2 \times C_2 \times C_2$ Lie groups:

There are three order eight finite groups. Of these, C_8 and $C_4 \times C_2$ hold only commutative spinor spaces and so hold only commutative Lie groups.

The order eight group $C_2 \times C_2 \times C_2$ holds five non-isomorphic spinor spaces. Of these five, two are commutative leaving only three non-commutative spinor spaces.

The group order eight $C_2 \times C_2 \times C_2$ has seven order four sub-groups and so its spinor spaces contain seven 4-dimensional spinor sub-spaces. Every imaginary variable is within a 4-dimensional spinor space. There are four types of 4-dimensional spinor space; two of these types of 4-dimensional spinor space are the commutative A_1 & A_2 algebras; the other two types of the 4-dimensional spinor spaces are the quaternion algebras and the A_3 algebras. Is one of the three 8-dimensional non-commutative spinor spaces such that it has seven non-commutative 4-dimensional sub-spaces?

There are three non-commutative $C_2 \times C_2 \times C_2$ spinor spaces to consider:

$$1 + 3\sqrt[2]{+1} + 4\sqrt[2]{-1}_{Non-Com}$$
$$1 + 5\sqrt[2]{+1} + 2\sqrt[2]{-1} \qquad (13.6)$$
$$1 + \sqrt[2]{+1} + 6\sqrt[2]{-1}$$

The *Non – com* subscript is given to this algebra because there is a commutative $1 + 3\sqrt[2]{+1} + 4\sqrt[2]{-1}_{Com}$ spinor space within the $C_2 \times C_2 \times C_2$ group.

The $1 + 5\sqrt[2]{+1} + 2\sqrt[2]{-1}$ algebra has the following 4-dimensional sub-algebras:

4	off	A_3	Non – Com	
2	off	A_1	Com	(13.7)
1	off	A_2	Com	

The $1 + \sqrt[2]{+1} + 6\sqrt[2]{-1}$ algebra has the following 4-dimensional sub-algebras:

4	off	$\mathbb{H}$	Non – Com	
3	off	A_2	Com	(13.8)

The $1 + 3\sqrt[2]{+1} + 4\sqrt[2]{-1}_{Non-Com}$ algebra has the following 4-dimensional sub-algebras:

1	*off*	$\mathbb{H}$	*Non – Com*	
3	*off*	A_3	*Non – Com*	(13.9)
3	*off*	A_2	*Com*	

We see that each non-commutative $C_2 \times C_2 \times C_2$ algebra has four non-commutative 4-dimensional sub-algebras and three 4-dimensional commutative sub-algebras.

Examination of the non-commutative $C_2 \times C_2 \times C_2$ algebras shows that the e variable always commutes with every other variable. These three algebras are actually the three 8-dimensional Clifford algebras in which the tri-vector $\overrightarrow{e_{123}}$, which is the e variable, is always fully commutative. The three commutative 4-dimensional sub-algebras are the three order four sub-groups which have the e variable as an element. There are no 4-dimensional $C_2 \times C_2$ non-commutative algebras which have a commutative imaginary variable, and so the 4-dimensional sub-algebras with the e variable must be commutative. We have six non-commutative variables, and so, within the non-commutative $C_2 \times C_2 \times C_2$ algebras, we have only six non-commutative generators for a Lie algebra.

Nor are all the non-commutative generators mutually non-commutative. Since we have three commutative 4-dimensional sub-algebras, we obviously have instances of two generators commuting with each other even though they might both non-commute with other variables.

The $1 + 3\sqrt[2]{+1} + 4\sqrt[2]{-1}_{Non-Com}$ algebras give rise to a 4-dimensional geometric space if the values of the variables are such that they form pairs. Since this is the only connection to our geometric 4-dimensional space-time, we will list the commutation relations of a particular example of this algebra.

Firstly, the $A \& E$ variables commute with every other variable. We will not list the commutation relations of these variables; nor will we list the other commutative relations. Unsurprisingly, the commutation relations fit into four sets corresponding to the four non-commutative 4-dimensional sub-algebras. We have:

$$
\begin{array}{lll}
[B,C] = -D, & [B,D] = C, & [C,D] = -B \\
[B,G] = H, & [B,H] = -G, & [G,\mathrm{H}] = B \\
[C,F] = H, & [C,H] = F, & [F,H] = C \\
[D,F] = G, & [F,G] = -D, & [D,G] = F
\end{array}
\quad (13.10)
$$

Corresponding to the natures of the sub-algebras, we have a copy of $SU(2)$ and three copies of the A_3 commutation relations which together form the Lorentz group $SO(3,1)$. Can this be used instead of $SU(3)$ within quantum field theory? We do not yet know. We do know that the $C_2 \times C_2 \times C_2$ really exist under the new approach to mathematics whereas, it seems, $SU(3)$ does not really exist under the new approach to mathematics.

Other individual $1 + 3\sqrt[2]{+1} + 4\sqrt[2]{-1}_{Non-Com}$ algebras contain anti-quaternions and anti-A_3 algebras.

A look at $SU(3)$:

The Lie algebra of $SU(3)$ is taken to be the eight linearly independent zero-trace unitary matrices presented as the Gell-Mann matrices[48]:

$$\lambda_1 = \begin{bmatrix} 0 & 1 & 0 \\ 1 & 0 & 0 \\ 0 & 0 & 0 \end{bmatrix} \quad \lambda_2 = \begin{bmatrix} 0 & -i & 0 \\ i & 0 & 0 \\ 0 & 0 & 0 \end{bmatrix} \quad (13.11)$$

$$\lambda_3 = \begin{bmatrix} 1 & 0 & 0 \\ 0 & -1 & 0 \\ 0 & 0 & 0 \end{bmatrix} \quad \lambda_4 = \begin{bmatrix} 0 & 0 & 1 \\ 0 & 0 & 0 \\ 1 & 0 & 0 \end{bmatrix} \quad (13.12)$$

$$\lambda_5 = \begin{bmatrix} 0 & 0 & -i \\ 0 & 0 & 0 \\ i & 0 & 0 \end{bmatrix} \quad \lambda_6 = \begin{bmatrix} 0 & 0 & 0 \\ 0 & 0 & 1 \\ 0 & 1 & 0 \end{bmatrix} \quad (13.13)$$

$$\lambda_7 = \begin{bmatrix} 0 & 0 & 0 \\ 0 & 0 & -i \\ 0 & i & 0 \end{bmatrix} \quad \lambda_8 = \frac{1}{\sqrt[2]{3}} \begin{bmatrix} 1 & 0 & 0 \\ 0 & 1 & 0 \\ 0 & 0 & -2 \end{bmatrix} \quad (13.14)$$

All of the above matrices are symmetric matrices and thus have real eigenvalues and orthogonal eigenvectors. Is λ_2 really a symmetric matrix; it looks anti-symmetric. Remembering the 2×2 matrix form of i, we have the symmetric matrix:

---

[48] See Georgi chapter 7.

$$\lambda_2 = \begin{bmatrix} 0 & 0 & 0 & -1 & 0 & 0 \\ 0 & 0 & 1 & 0 & 0 & 0 \\ 0 & 1 & 0 & 0 & 0 & 0 \\ -1 & 0 & 0 & 0 & 0 & 0 \\ 0 & 0 & 0 & 0 & 0 & 0 \\ 0 & 0 & 0 & 0 & 0 & 0 \end{bmatrix} \qquad (13.15)$$

We immediately see that the matrices $\{\lambda_1,\ \lambda_2,\ \lambda_3\}$ are the Pauli matrices we associate with $SU(2)$ set in a larger matrix, and these do indeed form an $SU(2)$ sub-group of $SU(3)$. The commutator of any two of these matrices is of the form of the third of these matrices multiplied by a scalar. In terms of their forms, these three matrices are a multiplicatively closed set. Each of them squares to 'almost the identity matrix' - an extra zero is added to the matrix to increase its size. In short, the three matrices, $\{\lambda_1,\ \lambda_2,\ \lambda_3\}$, and the unit matrix are 'like' an order four group written in obscure notation. As we saw above, to have a commutator Lie algebra, we need sets of three generators plus the identity.

The commutators of the above $SU(3)$ matrices are:

$$[\lambda_1 \quad \lambda_2] = 2i\lambda_3 \qquad [\lambda_1 \quad \lambda_3] = -2i\lambda_2$$
$$[\lambda_2 \quad \lambda_3] = 2i\lambda_1 \qquad (13.16)$$

$$[\lambda_1 \quad \lambda_4] = i\lambda_7 \qquad [\lambda_1 \quad \lambda_7] = -i\lambda_4$$
$$[\lambda_4 \quad \lambda_7] = i\lambda_1 \qquad (13.17)$$

$$\begin{bmatrix} \lambda_1 & \lambda_5 \end{bmatrix} = -i\lambda_6 \qquad \begin{bmatrix} \lambda_1 & \lambda_6 \end{bmatrix} = -i\lambda_5$$
$$\begin{bmatrix} \lambda_5 & \lambda_6 \end{bmatrix} = -i\lambda_1 \qquad (13.18)$$

$$\begin{bmatrix} \lambda_2 & \lambda_4 \end{bmatrix} = i\lambda_6 \qquad \begin{bmatrix} \lambda_2 & \lambda_6 \end{bmatrix} = -i\lambda_4$$
$$\begin{bmatrix} \lambda_4 & \lambda_6 \end{bmatrix} = i\lambda_2 \qquad (13.19)$$

$$\begin{bmatrix} \lambda_2 & \lambda_5 \end{bmatrix} = i\lambda_7 \qquad \begin{bmatrix} \lambda_2 & \lambda_7 \end{bmatrix} = -i\lambda_5$$
$$\begin{bmatrix} \lambda_5 & \lambda_7 \end{bmatrix} = i\lambda_2 \qquad (13.20)$$

Each of the matrices other than λ_8 squares to 'almost the identity matrix'. We see we have five multiplicatively closed sets of three matrices and the 'almost identity' matrix. Of course, there are only four non-commutative sub-algebras in each of the three non-isomorphic $C_2 \times C_2 \times C_2$ algebras, and so we have one too many sets of three matrices to match this group.

The other $SU(3)$ commutation relations are:

$$\begin{bmatrix} \lambda_3 & \lambda_4 \end{bmatrix} = i\lambda_5 \qquad \begin{bmatrix} \lambda_3 & \lambda_5 \end{bmatrix} = -i\lambda_4$$

$$\begin{bmatrix} \lambda_4 & \lambda_5 \end{bmatrix} = 2i \begin{bmatrix} 1 & 0 & 0 \\ 0 & 0 & 0 \\ 0 & 0 & -1 \end{bmatrix} \qquad (13.21)$$

$$\begin{bmatrix} \lambda_3 & \lambda_6 \end{bmatrix} = -i\lambda_7 \qquad \begin{bmatrix} \lambda_3 & \lambda_7 \end{bmatrix} = i\lambda_6$$

$$\begin{bmatrix} \lambda_6 & \lambda_7 \end{bmatrix} = 2i \begin{bmatrix} 0 & 0 & 0 \\ 0 & 1 & 0 \\ 0 & 0 & -1 \end{bmatrix} \qquad (13.22)$$

$$\begin{bmatrix} \lambda_4 & \lambda_8 \end{bmatrix} = -i\sqrt[2]{3}\lambda_5 \qquad \begin{bmatrix} \lambda_5 & \lambda_8 \end{bmatrix} = -i\sqrt[2]{3}\lambda_4 \qquad (13.23)$$

$$\begin{bmatrix} \lambda_5 & \lambda_8 \end{bmatrix} = i \sqrt[2]{3} \lambda_7 \qquad \begin{bmatrix} \lambda_7 & \lambda_8 \end{bmatrix} = i \sqrt[2]{3} \lambda_6 \qquad (13.24)$$

$$\begin{bmatrix} \lambda_1 & \lambda_8 \end{bmatrix} = 0 \quad \begin{bmatrix} \lambda_2 & \lambda_8 \end{bmatrix} = 0 \quad \begin{bmatrix} \lambda_3 & \lambda_8 \end{bmatrix} = 0 \quad (13.25)$$

We have two new matrices, and we have awkward $\sqrt[2]{3}$'s in some commutation relations.

If we multiply the $SU(3)$ matrices by $i = \sqrt[2]{-1}$, we get eight generators of 2-dimensional Euclidean rotations. It would seem that $SU(3)$ is a space with eight 2-dimensional rotations. Now, a 4-dimensional space has six ways of pairing two variables and a 5-dimensional space has ten ways of pairing two variables. There is no space which has eight ways of pairing two variables. Perhaps $SU(3)$, like $SU(2)$, is a spinor space. No, the only spinor spaces with as many 2-dimensional rotations as there are imaginary variables are the $C_2 \times C_2 \times ...$ spinor spaces; these spinor spaces have $(2^n - 1)$ 2-dimensional rotations where n is the number of C_2 groups crossed together. It does seem that $SU(3)$ is some kind of orphan.

Can we dump $SU(3)$?:

In general, to your author's tender eyes, $SU(3)$ looks too messy to be right. We had great success in extracting $U(1)$ and $SU(2)$ from the finite group algebras. Surely nature cannot be so ugly as $SU(3)$. However, we cannot dump $SU(3)$ until we have something with which to replace it. There have been successes with $SU(3)$; particles have been predicted using it. For the present, we must keep $SU(3)$, but we should be on the look-out for a replacement.

Sub-algebras of the $C_2 \times C_2 \times C_2 \times C_2$ algebras:

There are two non-isomorphic commutative 16-dimensional spinor algebras and five non-isomorphic non-commutative 16-dimensional spinor algebras within the $C_2 \times C_2 \times C_2 \times C_2$ group. The non-commutative algebras have the following sub-group structures.

The $1 + 9\sqrt[2]{+1} + 6\sqrt[2]{-1}$ algebra has:

6	A_1	sub-algebras
9	A_2	sub-algebras
18	A_3	sub-algebras
2	$\mathbb{H}$	sub-algebras

$$(13.26)$$

The $1 + 5\sqrt[2]{+1} + 10\sqrt[2]{-1}$ algebra has:

0	A_1	sub-algebras
15	A_2	sub-algebras
10	A_3	sub-algebras
10	$\mathbb{H}$	sub-algebras

$$(13.27)$$

The $1 + 3\sqrt[2]{+1} + 12\sqrt[2]{-1}$ algebra has:

1	A_1	sub-algebras
18	A_2	sub-algebras
0	A_3	sub-algebras
16	$\mathbb{H}$	sub-algebras

$$(13.28)$$

The $1 + 7\sqrt[2]{+1} + 8\sqrt[2]{-1}$ algebra has:

3	A_1	sub-algebras
15	A_2	sub-algebras
13	A_3	sub-algebras
4	$\mathbb{H}$	sub-algebras

$$(13.29)$$

The $1 + 11\sqrt[2]{+1} + 4\sqrt[2]{-1}$ algebra has:

13	A_1	sub-algebras
6	A_2	sub-algebras
16	A_3	sub-algebras
0	$\mathbb{H}$	sub-algebras

$$(13.30)$$

Chapter 14

The Other Order Eight Groups

The order eight dicyclic Lie group:

The Cayley table of the order eight dicyclic group is:

$$
\begin{bmatrix}
A_0 & A_1 & A_2 & A_3 & A_4 & A_5 & A_6 & A_7 \\
A_3 & A_0 & A_1 & A_2 & A_7 & A_4 & A_5 & A_6 \\
A_2 & A_3 & A_0 & A_1 & A_6 & A_7 & A_4 & A_5 \\
A_1 & A_2 & A_3 & A_0 & A_5 & A_6 & A_7 & A_4 \\
A_6 & A_5 & A_4 & A_7 & A_0 & A_3 & A_2 & A_1 \\
A_7 & A_6 & A_5 & A_4 & A_1 & A_0 & A_3 & A_2 \\
A_4 & A_7 & A_6 & A_5 & A_2 & A_1 & A_0 & A_3 \\
A_5 & A_4 & A_7 & A_6 & A_3 & A_2 & A_1 & A_0
\end{bmatrix}
$$

The order eight dicyclic group holds 128 copies of only one division algebra, spinor space. The algebra is non-commutative and is of the form $1+\sqrt[2]{+1}+6\sqrt[4]{+1}$. The rotation matrix of this algebra is the Lie group. We have one 2-dimensional rotation in this Lie group and six 4-dimensional rotations. There are seven generators of the Lie algebra. The commutation relations are:

$$
\begin{aligned}
&[C,B]=0, \ \ [C,D]=0, \ \ [C,E]=0 \\
&[C,F]=0, \ \ [C,G]=0, \ \ [C,H]=0
\end{aligned}
\tag{14.1}
$$

$$[B,D] = 0,$$
$$[B,E] = [B,G] = [D,E] = [D,G] = F + H \quad (14.2)$$
$$[B,F] = [B,H] = [D,F] = [D,H] = E + G$$
$$[E,G] = [F,H] = 0$$
$$[E,F] = [E,H] = [F,G] = [G,H] = B + D \quad (14.3)$$

We see we have no commutator which gives a single other generator. We also see that we have one generator, C, which commutes with all other variables.

The order eight dihedral group:

The Cayley table of the order eight dihedral group is:

$$
\begin{bmatrix}
A_0 & A_1 & A_2 & A_3 & A_4 & A_5 & A_6 & A_7 \\
A_3 & A_0 & A_1 & A_2 & A_5 & A_6 & A_7 & A_4 \\
A_2 & A_3 & A_0 & A_1 & A_6 & A_7 & A_4 & A_5 \\
A_1 & A_2 & A_3 & A_0 & A_7 & A_4 & A_5 & A_6 \\
A_4 & A_5 & A_6 & A_7 & A_0 & A_1 & A_2 & A_3 \\
A_5 & A_6 & A_7 & A_4 & A_3 & A_0 & A_1 & A_2 \\
A_6 & A_7 & A_4 & A_5 & A_2 & A_3 & A_0 & A_1 \\
A_7 & A_4 & A_5 & A_6 & A_1 & A_2 & A_3 & A_0
\end{bmatrix}
$$

The order eight dihedral group holds 64 copies of the $1 + 2\sqrt[4]{+1} + 3\sqrt[2]{+1} + 2\sqrt[2]{-1}$ algebra, it also holds 32 copies of the $1 + 2\sqrt[4]{+1} + 5\sqrt[2]{+1}$ algebra and it also holds 32 copies of the $1 + 2\sqrt[4]{+1} + \sqrt[2]{+1} + 4\sqrt[2]{-1}$ algebra.

The algebras are all non-commutative. The rotation matrix of these algebras are the Lie groups of these algebras. We have six 2-dimensional rotations in this Lie group and two 4-dimensional rotations. There are seven generators of the Lie algebra.

The commutation relations are:

$$[C,B] = 0, \quad [C,D] = 0, \quad [C,E] = 0$$
$$[C,F] = 0, \quad [C,G] = 0, \quad [C,H] = 0 \tag{14.4}$$

$$[B,D] = 0,$$
$$[B,E] = [B,G] = [D,E] = [D,G] = F + H \tag{14.5}$$
$$[B,F] = [B,H] = [D,F] = [D,H] = E + G$$

$$[E,G] = [F,H] = 0$$
$$[E,F] = [E,H] = [F,G] = [G,H] = B + D \tag{14.6}$$

The astute reader will notice these are an exact match for the dicyclic group (14.1) & (14.2) & (14.3). This is not a printing error. It is not understood why these two sets of commutation relations should be the same. Again the C variable commutes with all other variables.

Two commutative variables in 8-dim:
There are three order eight finite groups which hold non-commutative spinor algebras. Two of these three order eight

finite groups, the order eight dicyclic group and the order eight dihedral group are the non-commutative groups shown immediately above; the other of these three order eight finite groups is the commutative $C_2 \times C_2 \times C_2$ finite group. Remarkably, all three of these groups have commutation relations in which two variables, the real variable and one imaginary variable, commute with all other elements of the group. Technically, the groups have a centre of two elements. In the cases of the two non-commutative order eight groups, the C variable is the non-identity variable within the centre of the group. In the $C_2 \times C_2 \times C_2$ case, it is the E variable which is the non-identity variable within the centre.

Since the $C_2 \times C_2 \times C_2$ group algebras are the Clifford algebras and the E variable corresponds to the basis tri-vector $\overrightarrow{e_{123}}$, we understand the commutativity of this element. We do not properly understand why all the other non-commutative spinor algebras of order eight should similarly have a commutative element.

Intriguingly, there are two commutative Casimir elements associated with $SU(3)$.

Chapter 15

Review of Lie Theory

There has been much material presented lately. In this short chapter, we review and restate what has been presented without the details. In the following chapters, we will look at the conventional approach to Lie theory a little more.

Two types of Lie group:
There are two types of space, spinor space and geometric space, and so there are two types of Lie group, continuous group. Each Lie group is associated with a Lie algebra, and so there are two types of Lie algebra.

Spinor space Lie algebras:
We take a finite group written as a set of permutation matrices. We form the algebraic matrix form of this finite group and thus, by scattering minus signs throughout the algebraic matrix form, all possible division algebra matrix forms of the finite group. We take the exponential of the various algebraic matrix forms to get the polar form of each algebra. The rotation matrix is a continuous spinor group, a spinor Lie group.

To meet the conventional Lie algebra associated with the Lie group, for each algebra, we take the separate imaginary variable matrices of the Cartesian form of the algebra and set the variable to one. The resulting set of matrices are the generators which are the elements of the Lie algebra. We might have to multiply some of these by $i = \sqrt[2]{-1}$ to convert them into matrices with real eigenvalues. If the spinor algebra is non-commutative, we impose the commutation bracket upon the generator matrices and we have the corresponding Lie algebra.

When the spinor algebra is commutative, we still have the generators, but without the commutation bracket, this is a 'commutative Lie algebra'. Commutative Lie algebras are not 'really' Lie algebras in the conventional view; other than $U(1)$, they seem to play little part in physics.

When the commutation bracket gives something other than a single generator, as we saw it do in the case of the finite group S_3, (13.4) & (13.5), then the conventional view is that this is not a Lie algebra.

Thus, we get the Lie algebra of a Lie group by taking the unit imaginary variables of the Cartesian form of a spinor algebra.

If the finite group is such that each imaginary variable forms a C_2 sub-group with the identity (these are only the $C_2 \times C_2 \times ...$ groups), then the generators will generate a 2-dimensional rotation matrix of either Euclidean form or space-time form.

Conventional views of Lie algebra are based on the assumption that all types of space have 2-dimensional planes formed by every pair of axial variables. Such spaces are geometric spaces. In such spaces, we can rotate 2-

dimensionally in every 2-dimensional plane. Constrained by this conventional view of space, conventional Lie algebras are associated with only the $C_2 \times C_2 \times \ldots$ groups.

We might have to write the generator matrices in a different notation or in a different basis to conform with the conventional expression of the Lie algebra as we did above in the $SU(2)$ case.

Geometric space Lie algebras:

Geometric spaces have Lie groups represented by more than one rotation matrix. Each rotation matrix is a 2-dimensional rotation matrix set in a larger matrix. The generator of such a matrix is the matrix whose exponential gives the rotation matrix. Such Lie groups have as many generators as there are pairs of variables in the distance function respected by the Lie group. When we add the commuter bracket to the set of generator matrices, we have the Lie algebra of the particular geometric space.

We might have to write the generator matrices in a different notation or in a different basis to conform with the conventional expression of the Lie algebra.

The conventional approach:

Conventionally, a Lie algebra is seen as a set of matrices with particular properties and particular relations between them. The names given to the Lie groups and Lie algebras reflect this 'type of matrix' view. The Lie group $U(1)$ is the set of 1×1 unitary matrices with unit determinant. The Lie group $SU(2)$ is the set of 2×2 traceless unitary matrices with determinant unity. The Lie group $SO(3)$ is the set of 3×3 traceless orthogonal matrices with determinant unity.

The conventional approach is to seek all such unitary matrices and to call all these unitary matrices the Lie algebra. Such a procedure will lead to the conventional Lie group $SU(3)$. Since we do not find $SU(3)$ within the spinor spaces, there is clearly a conflict here. The clearly extant conflict is not clearly understood[49].

---

[49] I think I've made that clear.

Chapter 16

Orthogonal Groups

In these last chapters, we tidy a little of the conventional nomenclature. We include these chapters for completeness rather than because there is any great insight.

There is a tendency within conventional mathematics to presume that there exist geometric spaces of every dimension and of a purely Euclidean nature. These spaces are called Riemann or Euclidean spaces and have distance functions which are (all pluses) quadratic forms like:

$$dist^2 = a^2 + b^2$$
$$dist^2 = a^2 + b^2 + c^2 + d^2 + e^2 + ...$$
(16.1)

The only spaces with these distance functions which derive from the finite groups are the 2-dimensional complex plane, $\mathbb{C}$, the two 4-dimensional quaternion spaces, the 4-dimensional expectation space of the quaternions and its sub-spaces, and the 3-dimensional geometric sub-space of the A_3 expectation space and the 2-dimensional geometric sub-space of the A_3 expectation space. From the finite groups, there emerge no spaces of dimension more than four which have an 'all pluses' Riemann, Euclidean, distance function.

Within any set of $n \times n$ matrices, there are some matrices which preserve the Euclidean distance function of n-dimensional Euclidean space as measured by the length of a real vector acted upon by the matrix. In other words, we multiply a n-component real vector by the matrix and the norm of the vector, measured using a Euclidean distance function, is unchanged by the multiplication.

Such matrices are either rotation matrices with a particular angle or are reflection matrices. Such matrices are a group because when one such matrix is acted upon by another such matrix we get a third such matrix and each such matrix has an inverse and the identity matrix is one of the group. For any particular n, this group of $n \times n$ matrices is called the orthogonal group and is written as $O(n)$.

The reflection matrices have determinant minus unity; the rotation matrices have determinant plus unity. If we exclude the reflection matrices with determinant minus unity, then the group of these Euclidean distance preserving matrices with determinant plus unity is called the special orthogonal group and is written as $SO(n)$.

We have defined special orthogonal matrices as matrices which preserve the Euclidean distance function but are not reflections. These are just rotations. In 2-dimensional space, we have the whole of $SO(2)$ written as the single rotation matrix:

$$SO(2) = \begin{bmatrix} \cos\theta & \sin\theta \\ -\sin\theta & \cos\theta \end{bmatrix} \qquad (16.2)$$

At this point, the reader might be wondering what all the fuss is about. Quite so! There is confusion around this matter due to a lack of understanding of the difference between geometric space and spinor space.

When a mathematician decides to investigate a Riemann space of, say, five dimensions, he assumes that there are as many angles in the space as there are 2-dimensional planes in the space. From where come all these angles? Only spinor spaces have angles. This assumption of the magical appearance of just the right number of angles is effectively to assume the Riemann space is a geometric space. Having so assumed the existence of a Euclidean space of the appropriate dimension, we discover that there are sets of matrices which are effectively the 2-dimensional, (16.2), rotation matrix with a few 1's scattered around. We have seen an example of this above when we looked at $SO(3)$, (8.2).

These $n \times n$ matrices with $SO(2)$ in the various corners are the special orthogonal groups of the appropriate dimension. Underlying the idea that these distance preserving matrices form a group is the idea that a Euclidean space of the appropriate dimension exists and that it is a geometric space in which we can wave our arms around because we can rotate in every 2-dimensional plane within the space. If the space does not exist, and, within the new approach to mathematics, it does not for dimension higher than four, then the distance in that space cannot be preserved and we are telling fairy tales when we speak of special orthogonal groups of dimension greater than four.

The reader might have noticed that the quaternion rotation matrix, (9.1), is a group which preserves the 4-dimensional Euclidean distance function. Surely this is a member of the

4-dimensional special orthogonal group $SO(4)$. No, the quaternion rotation matrix preserves the Euclidean distance function in a 4-dimensional spinor space; this is $SU(2)$. The definition of the 4-dimensional special orthogonal groups does not exclude $SU(2)$ because it is a poor definition. Your author excludes $SU(2)$ from the 4-dimensional special orthogonal group because it is within an entirely different type of space from the geometric spaces of the special orthogonal groups.

So that's it. Orthogonal groups, special or general, are associated with geometric spaces in which we can wave our arms around. We will see that another type of matrix group, unitary groups, is associated with spinor spaces. In the 2-dimensional case, the spinor space coincides with the geometric space, and so the unitary group $U(1)$ is isomorphic to the special orthogonal group $SO(2)$.

There is another way of defining orthogonal matrices. An orthogonal matrix is a real matrix whose inverse is equal to its transpose:

$$M^{-1} = M^{T} \tag{16.3}$$

This definition also includes the quaternion rotation matrix, but we will not include the quaternion matrix because it exists in a spinor space.

By accident[50], in only Euclidean spaces, the two definitions of orthogonal matrices coincide.

---

[50] Of course, nothing happens by accident in mathematics.

Note: There is a rumour moving through the lecture rooms of the mathematics faculties of our universities that a matrix is not a rotation matrix unless it is orthogonal. Although this is true in Euclidean spaces, it is nonsense in other spaces.

Chapter 17

Unitary Groups

The unitary group of dimension n is the group of $n \times n$ unitary matrices. A matrix is unitary if its conjugate transpose is equal to its inverse. Unitary matrices have determinant plus or minus unity and they preserve the Euclidean distance function. The reader might realise that unitary matrices are very much like orthogonal matrices. Conventionally, unitary matrices are basically orthogonal matrices with complex numbers for elements rather than real numbers.

Within this book, we like to take the view that orthogonal groups preserve the distance function in geometric spaces but it is unitary groups which preserve the distance in spinor spaces. There are only two spinor spaces which have Euclidean distance functions, the complex numbers, $\mathbb{C}$, and the quaternion algebras, and so, this book is of the view that there exist only two unitary groups, $U(1)$ and $SU(2)$. This is a very different view from the conventional view. For a start, we do not have $SU(3)$.

An example of a unitary matrix is:

$$\begin{bmatrix} a+ib & c+id \\ -c+id & a-ib \end{bmatrix} \; : \; a^2 + b^2 + c^2 + d^2 = 1 \quad (17.1)$$

The conjugate transpose of this is:

$$\begin{bmatrix} a-ib & -c-id \\ c-id & a+ib \end{bmatrix} \quad : \quad a^2+b^2+c^2+d^2=1 \quad (17.2)$$

The product of these, (17.1) & (17.2), will be the identity because $a^2+b^2+c^2+d^2=1$.

Hermitian matrices:

Hermitian matrices are matrices whose conjugate transpose is equal to themselves:

$$A = \left(A^T\right)^* \quad (17.3)$$

Examples are:

$$\begin{bmatrix} x & a+ib \\ a-ib & x \end{bmatrix} \equiv \begin{bmatrix} x & 0 & a & b \\ 0 & x & -b & a \\ a & -b & x & 0 \\ b & a & 0 & x \end{bmatrix} \quad (17.4)$$

And:

$$\begin{bmatrix} x & a+ib & c+id \\ a-ib & x & e+if \\ c-id & e-if & x \end{bmatrix} \equiv \begin{bmatrix} x & 0 & a & b & c & d \\ 0 & x & -b & a & -d & c \\ a & -b & x & 0 & e & f \\ b & a & 0 & x & -f & e \\ c & -d & e & -f & x & 0 \\ d & c & f & e & 0 & x \end{bmatrix}$$

$$(17.5)$$

Physicists are particularly interested in hermitian matrices because these matrices have real eigenvalues and orthogonal eigenvectors. Hermitian matrices are just symmetric matrices in a less clear notation. We have met many symmetric matrices within the $C_2 \times C_2 \times ...$ groups. We have met anti-symmetric matrices in these groups also. Anti-symmetric matrices have only imaginary eigenvalues (they come in conjugate pairs with a zero if needed). We can make an anti-symmetric matrix into a matrix with real eigenvalues by multiplying it by $i = \sqrt[2]{-1}$.

Hermitian matrices are just symmetric matrices with a zero variable 'attached to' the real variable. Hermitian matrices have real eigenvalues and orthogonal eigenvectors; so do symmetric matrices.

The eigenvalues of (17.4) are respectively:

$$\begin{Bmatrix} x + \sqrt[2]{a^2 + b^2} \\ x - \sqrt[2]{a^2 + b^2} \end{Bmatrix} \quad \& \quad \begin{Bmatrix} x + \sqrt[2]{a^2 + b^2} \\ x - \sqrt[2]{a^2 + b^2} \\ x + \sqrt[2]{a^2 + b^2} \\ x - \sqrt[2]{a^2 + b^2} \end{Bmatrix} \qquad (17.6)$$

The eigenvectors of (17.4) are:

$$\begin{bmatrix} \dfrac{x}{\kappa} + a + ib \\ -\dfrac{x}{\kappa} + a + ib \end{bmatrix} \quad \& \quad \begin{bmatrix} x - \kappa \\ x - \kappa \end{bmatrix} \qquad (17.7)$$

$$\kappa = \sqrt[2]{a^2 + b^2}$$

133

And:

$$
\begin{bmatrix} \dfrac{x}{\kappa}+b \\[6pt] \dfrac{x}{\kappa}+a \\[6pt] \dfrac{x}{\kappa}-b \\[6pt] \dfrac{x}{\kappa}-a \end{bmatrix}
\quad \& \quad
\begin{bmatrix} \dfrac{x}{\kappa}+a \\[6pt] \dfrac{x}{\kappa}-b \\[6pt] \dfrac{x}{\kappa}-a \\[6pt] \dfrac{x}{\kappa}+b \end{bmatrix}
\quad \& \quad
\begin{bmatrix} 0 \\[6pt] x-\kappa \\[6pt] 0 \\[6pt] x-\kappa \end{bmatrix}
\quad \& \quad
\begin{bmatrix} x-\kappa \\[6pt] 0 \\[6pt] x-\kappa \\[6pt] 0 \end{bmatrix}
\qquad (17.8)
$$

Putting it succinctly, hermitian matrices are just symmetric matrices in a different notation. Of course, within the various different algebras of the different ordered $C_2 \times C_2 \times \ldots$ groups, we have various symmetric variables. These symmetric variables are hermitian generators.

Chapter 18

Representations Briefly

Within physics, an important aspect of Lie algebra is the representations of a Lie algebra (not a Lie group). We briefly introduce the reader to this concept.

Although we think of a Lie algebra as being a set of generators of rotations with a commutation operation, convention takes the essence of the Lie algebra to be the actual commutation relations. The generators are matrices that hold, apart from a scalar multiple, the commutation relations. Having obtained the generators, and thus the commutation relations, convention takes a step away from the Lie group and declares that the Lie algebra can be presented as any set of matrices, formally called linear transformations, which have the same commutation relations.

For example, we can represent the commutation relations of $SU(2)$ as the three 2×2 matrices[51]:

$$J_1^{1/2} = \frac{1}{2}\begin{bmatrix} 0 & 1 \\ 1 & 0 \end{bmatrix}, \quad J_2^{1/2} = \frac{1}{2}\begin{bmatrix} 0 & -i \\ i & 0 \end{bmatrix}$$

$$J_3^{1/2} = \frac{1}{2}\begin{bmatrix} 1 & 0 \\ 0 & -1 \end{bmatrix} \tag{18.1}$$

---

[51] See: Georgi Chapter 3 page 61

Representations Briefly

We can represent the same set of commutation relations as the three 3×3 matrices:

$$J_1^1 = \frac{1}{\sqrt[2]{2}}\begin{bmatrix} 0 & 1 & 0 \\ 1 & 0 & 1 \\ 0 & 1 & 0 \end{bmatrix}, \quad J_2^1 = \frac{1}{\sqrt[2]{2}}\begin{bmatrix} 0 & -i & 0 \\ i & 0 & -i \\ 0 & i & 0 \end{bmatrix}$$

$$J_3^1 = \frac{1}{\sqrt[2]{2}}\begin{bmatrix} 1 & 0 & 0 \\ 0 & 0 & 0 \\ 0 & 0 & -1 \end{bmatrix} \tag{18.2}$$

For example:

$$\begin{bmatrix} J_1^{1/2}, & J_2^{1/2} \end{bmatrix} = iJ_3^{1/2}$$
$$\begin{bmatrix} J_1^1, & J_2^1 \end{bmatrix} = iJ_3^1 \tag{18.3}$$

The first of these sets of matrices, (18.1), is called the spin-half representation of $SU(2)$, and the second of these sets of matrices, (18.2), is called the spin-one representation of $SU(2)$.

Of course, the decision to use complex numbers within the matrices effectively halves the size of the matrices from, in the case of (18.1), 4×4 matrices to 2×2 matrices.

The idea of having differently sized matrices representing the same set of commutation relations is that they can act upon different sized vectors. For example, the 2×2 matrices, can act on a two-component complex vector:

$$\begin{bmatrix} a+ib \\ c+id \end{bmatrix} \tag{18.4}$$

136

The 3×3 matrices, (18.2), can act on a three-component complex vector. If we use only real elements in the matrices, thereby doubling the size of the matrix, the size of the vector doubles also.

We know that a pair of complex numbers is a way of writing a quaternion, and so the 2×2 matrices, (18.1), when written as 4×4 matrices, act on a 4×4 quaternion matrix. There is a problem in that the 2×2 matrices, (18.1), even when written as 4×4 matrices, are not quaternion matrices. If we multiply the 2×2 matrices, (18.1), by i, they become quaternion matrices. Thus, after ducking and diving around, we come to a quaternion equation – we are within a proper division algebra. The 2×2 matrices, (18.1), and their 4×4 equivalents, are associated with rotation because they are generators of 2-dimensional rotations. We thus convince ourselves that the vector (quaternion) upon which the 4×4 matrices act is rotated by their action – phew!.

We see the rotation is far from neat and clean. For this reason, physicists often do not speak of rotations but speak of transformations instead.

On top of all this, we have the vector of three complex components acted upon by the 3×3 matrices, (18.2). What on Earth is this vector? It is certainly not a quaternion. There are representations of $SU(2)$, sets of matrices, of all sizes. Upon what do they act?

Representations are not nonsense:
Looking at the above, we might wonder why we are interested in representations of Lie algebras at all. Surely,

our interest in in the Lie groups. Remarkably, representations seem to play a role in physics. Representations are used by physicists because they give useful results not because physicists understand what is really happening.

Reducible and irreducible representations:

There are times in physics when we wish to consider a number of vectors together or we wish to tie two rotations together. We might form a rotation matrix like:

$$\begin{bmatrix} \left[\mathbb{H}_{rot}\right] & 0 \\ 0 & \left[\mathbb{H}_{rot}\right] \end{bmatrix} \tag{18.5}$$

This is just two 4×4 quaternion rotation matrices on the leading diagonal of a 8×8 matrix. They act independently on a pair of quaternions written as a column vector. We might sensibly take the view that the 8×8 matrix could be split into two separate 4×4 matrices. Such a rotation as (18.5) is said to be reducible. A rotation matrix which cannot be presented as blocks on the leading diagonal of a larger matrix is said to be irreducible.

Representations, like rotations, can be collected together and placed on the leading diagonal like (18.5). Such representations are said to be reducible. It is the irreducible representations which interest physicists.

Remarkably, and we know not whether this is of any import, the $C_2 \times C_2 \times C_2$ group contains algebras similar to (18.5) like:

$$\begin{bmatrix} \mathbb{H}_1 & \mathbb{H}_2 \\ \mathbb{H}_2 & \mathbb{H}_1 \end{bmatrix} \qquad (18.6)$$

Setting four of the variables to zero will give:

$$\begin{bmatrix} \mathbb{H}_1 & 0 \\ 0 & \mathbb{H}_1 \end{bmatrix} \qquad (18.7)$$

Unitary representations:

A unitary representation of a group is a set of matrices (linear transformations) whose multiplicative relations copy the multiplicative relations of the elements of the group they represent and whose determinants are all plus or minus one.

Chapter 19

How to get into a Muddle

In this chapter, we follow the book 'An Elementary Primer for Gauge Theory' by K. Moriyasu[52]. We are interested in the appendix of this book page 158. Your author has a very high opinion of this book.[53] This book has been chosen because of the clarity of exposition not because of any muddled exposition. Moriyasu's book presents Lie algebras as clearly as they can be presented, and this allows your author to most easily untangle the muddle of Lie algebra.

Moriyasu begins with a clear description of the continuous group $U(1)$ as the circle in the complex plane. Then we begin to get muddled.

Muddle step 1: Instead of writing a rotation matrix as a matrix:

$$U(\theta) = \begin{bmatrix} \cos\theta & \sin\theta \\ -\sin\theta & \cos\theta \end{bmatrix} \qquad (19.1)$$

use concise notation:

$$U(\theta) = e^{i\theta} \qquad (19.2)$$

---

[52] ISBN: 9971-950-83-9

[53] It was given to your author by a friend, and it was your author's first introduction to gauge theory.

Concise notation has a place in mathematics. It is very useful in dealing with infinite matrices for example. Properly used, concise notation can display the links between different parts of a mathematical structure. Unfortunately, in your author's opinion, concise notation is most often used because it is fashionable or, worse, to impress the reader with the writer's mathematical prowess.

Moriyasu goes on to show that $U(1)$ satisfies the group axioms of closure and inverse:

$$U(\theta)U(\theta') = U(\theta+\theta')$$
$$U^{-1}(\theta) = U(-\theta)$$
(19.3)

It would have been easier to simply say that 'two rotations make a rotation and, for every rotation, there is an inverse rotation. This, (19.3), is basically just a few trigonometric identities.

Moriyasu then shows that $U(1)$ is differentiable; by this we mean that the derivative of $U(\theta)$ is an element of the group $U(1)$. We have:

$$dU = U(\theta+d\theta) - U(\theta)$$
$$= e^{i\theta}(1+id\theta) - e^{i\theta}$$
$$= ie^{i\theta}d\theta = iUd\theta$$
(19.4)

It would have been easier to simply differentiate the matrix with respect to the imaginary variable. We use the imaginary variable because the rotation matrix is the exponential of the imaginary variable:

$$\exp \begin{bmatrix} 0 & \theta \\ -\theta & 0 \end{bmatrix} = \begin{bmatrix} \cos\theta & \sin\theta \\ -\sin\theta & \cos\theta \end{bmatrix} \tag{19.5}$$

We have:

$$\frac{\partial \begin{bmatrix} \cos\theta & \sin\theta \\ -\sin\theta & \cos\theta \end{bmatrix}}{\partial \begin{bmatrix} 0 & \theta \\ -\theta & 0 \end{bmatrix}} = \frac{1}{\begin{bmatrix} 0 & 1 \\ -1 & 0 \end{bmatrix}} \frac{\partial \begin{bmatrix} \cos\theta & \sin\theta \\ -\sin\theta & \cos\theta \end{bmatrix}}{\partial \begin{bmatrix} \theta & 0 \\ 0 & \theta \end{bmatrix}}$$

$$= \begin{bmatrix} 0 & -1 \\ 1 & 0 \end{bmatrix} \begin{bmatrix} -\sin\theta & \cos\theta \\ -\cos\theta & -\sin\theta \end{bmatrix} \tag{19.6}$$

$$= \begin{bmatrix} \cos\theta & \sin\theta \\ -\sin\theta & \cos\theta \end{bmatrix}$$

There is an implication to $U(1)$ being differentiable as is pointed out by Moriyasu, "*The distinguishing characteristic of a Lie group is that the parameters of a product must be analytic functions of the parameters of each factor in the product.*" This is essential if the group is to be differentiable. Conventional Lie algebra uses the infinitesimal dU; we need the group to be differentiable for this infinitesimal to exist. Wow! Why not just say that the rotation matrix is an element of a division algebra such as the complex numbers?

There is necessity within mathematics for precise technical definition, but without the intuitive understanding, it is easy to get lost in a tangle of technicalities.

Moriyasu explains compactness very well: "*A compact Lie group is one for which the parameters are allowed to range over only a closed interval. The group $U(1)$ is compact*

because the angle θ is defined over the interval $[0, 2\pi]$*. ...*
The property of compactness guarantees that the group is
unitary."

Unitary means to do with circles or spheres rather than with hyperbolas. Essentially, unitary groups are rotations in spaces with 'all pluses' quadratic distance functions of the form:

$$d^2 = a^2 + b^2 + c^2 + ... + z^2 \qquad (19.7)$$

Such spaces are called Riemann spaces. Non-unitary groups have at least one minus sign in the quadratic distance function, like our space-time; these are called pseudo-Riemann spaces. Non-unitary groups are non-compact. So there we have it; unitary and compact are the same thing; unitary is Euclidean space; non-unitary is non-compact is hyperbolic space-time. Unitary groups are rotations in Euclidean spaces. Non-unitary groups are rotations in non-euclidean spaces.[54] The Lorentz group $SO(3,1)$ is non-compact.

Unitarity is important:

Moriyasu goes on to explain that unitary groups are important because they preserve the modulus squared of a complex wave function. Within quantum mechanics, the probability of anything happening is the modulus squared of

---

[54] The only division algebras which have unitary distance functions are the complex numbers, $\mathbb{C}$, and the two quaternion algebras – no 8-dimensional division algebra has a unitary distance function.

the wave function. The wave function has two forms; a single complex number field, and a pair of complex numbers field (spinor field). These are:

$$|\psi|^2 = a^2 + b^2$$
$$|\psi|^2 = a^2 + b^2 + c^2 + d^2$$

(19.8)

Unitary Lie groups are of such interest to physicists precisely because rotation in a Euclidean space, which is what a unitary Lie group is, preserves the Euclidean distance function which is exactly (19.8). In other words, the probability of something happening does not change with rotation in euclidean space – an observer looking west sees the same physics as an observer looking north.

How to get really muddled:

Moriyasu now goes on to look at $SU(2)$. We emphasize that the muddle here is not of Moriyasu's making but is the conventional approach to Lie groups.

We begin by forming four real numbers into a two-component complex vector (spinor), and we seek a rotation matrix which will rotate that complex vector. We need a 2×2 rotation matrix to rotate a two-component vector:

$$\begin{bmatrix} u' \\ v' \end{bmatrix} = \begin{bmatrix} a & b \\ c & d \end{bmatrix} \begin{bmatrix} u \\ v \end{bmatrix}$$

(19.9)

Muddle step 2: Why on Earth do we use a complex vector rather than a 4-component real vector? It is almost as if this is done deliberately to confuse the student. The reason is an historical accident. In the 1920's and 1930's, and still today, physicists knew little or nothing of quaternions. Only a madman would write a quaternion as two complex numbers, but we are all mad.

Having invented complex vectors, we invent spaces with complex axes in which these complex vectors exist, $\mathbb{C}^n$. We wrongly add a complex inner product[55] and call this space a complex Hilbert space. Complex axes – insane. By accident, the inner product works because a pair of complex numbers with the correct norm is a quaternion. By this accident, generations have been confused into thinking that complex Hilbert spaces really exist.

Having got a two-component vector, we require the probability, $P = |u|^2 + |v|^2$, to be preserved and, since the 2×2 matrix is a rotation matrix, we require the determinant of the 2×2 matrix to be unity. This drives us to that matrix:

$$\begin{bmatrix} a & b \\ -b^* & a^* \end{bmatrix} \quad : \quad \{a,b\} \in \mathbb{C} \quad (19.10)$$

This, (19.10), is a quaternion, but it does not look like it. The condition that the determinant is unity makes (19.10) the quaternion rotation matrix.

We can now identify this matrix, (19.10), with a rotation matrix in 2-dimensional complex space; it is really a rotation matrix in 4-dimensional quaternion space. We call this matrix, (19.10), $SU(2)$. We choose the 2 because the

---

[55] Inner products exist in only division algebras.

matrix is 2×2, and we say this is the special unitary group of (complex) dimension two.

Muddle step 3: We notice that we can associate the rotation matrix in 2-dimensional complex space, $\mathbb{C}^2$, with rotation in the 3-dimensional real sub-space of our 4-dimensional space-time, $\mathbb{R}^3$, by putting:

$$
\begin{aligned}
x &= \frac{1}{2}\left(u^2 - v^2\right) \\
y &= \frac{1}{2i}\left(u^2 + v^2\right) \\
z &= uv
\end{aligned}
\tag{19.11}
$$

Calculation shows that the $SU(2)$ rotation matrix, (19.10), which, by-the-way, does not look anything like a rotation matrix, will preserve the 3-dimensional distance function $d^2 = x^2 + y^2 + z^2$ as well as preserving the distance function in 2-dimensional complex space. Thus we have one rotation matrix for two completely different spaces. Is the reader muddled enough yet because there's worse to come?

We discover that in order to apply the $SU(2)$ rotation matrix to 3-dimensional real space, we need to use half-angles[56]:

---

[56] Thank you Moiyasu.

$$SU(2) = \begin{bmatrix} \cos\dfrac{\beta}{2}e^{i\frac{(\alpha+\gamma)}{2}} & \sin\dfrac{\beta}{2}e^{-i\frac{(\alpha+\gamma)}{2}} \\[3mm] -\sin\dfrac{\beta}{2}e^{i\frac{(\alpha+\gamma)}{2}} & \cos\dfrac{\beta}{2}e^{-i\frac{(\alpha+\gamma)}{2}} \end{bmatrix} \quad (19.12)$$

We say that $SU(2)$ double covers rotation in $\mathbb{R}^3$ because the half angles imply we need to rotate through 720^0 to get back to from where we started rather than through 360^0.

Group generators:

Well, by now, we are in a right muddle. Let us recall that $SU(2)$ is just the quaternion rotation matrix and that it acts on quaternions. Quaternion rotation is nothing to do with rotation in our 4-dimensional space-time or with any sub-space of our 4-dimensional space-time. I feel better now.

Based upon the assumption that rotation within quaternion space, masquerading as 2-dimensional complex space, is 2-dimensional like rotation is in our space-time, we divide the $SU(2)$ rotation matrix into separate parts, one for each 2-dimensional rotational plane.

In quaternion form, these parts would be:

$$\begin{bmatrix} 0 & \theta & 0 & 0 \\ -\theta & 0 & 0 & 0 \\ 0 & 0 & 0 & -\theta \\ 0 & 0 & \theta & 0 \end{bmatrix}, \begin{bmatrix} 0 & 0 & \phi & 0 \\ 0 & 0 & 0 & \phi \\ -\phi & 0 & 0 & 0 \\ 0 & -\phi & 0 & 0 \end{bmatrix}, \begin{bmatrix} 0 & 0 & 0 & \varphi \\ 0 & 0 & -\varphi & 0 \\ 0 & \varphi & 0 & 0 \\ -\varphi & 0 & 0 & 0 \end{bmatrix}$$

$$(19.13)$$

In quaternion form, we would simply add these parts and take the exponential of the sum to get the quaternion rotation matrix. In the muddled form, we represent these parts as three 2×2 hermitian (self-adjoint) matrices.

There are only three linearly independent hermitian 2×2 matrices, and so we call these the $SU(2)$ generators, and we say these three $SU(2)$ generators are the Lie algebra $SU(2)$.

We discover that the three linearly independent hermitian 2×2 matrices, the $SU(2)$ generators, are non-commutative. Of course they are, quaternions are non-commutative. We write out the commutation relations of these matrices; they are the same commutation relations as the quaternions; and we say this set of commutation relations are the Lie algebra $SU(2)$. Back to Moriyasu. *"The generators have a mathematical structure that is very different from the group itself. The group elements are multiplied together, while the generators are added."* Yes! the generators are quaternion variables which are added to form the complete quaternion. The group elements are just quaternion rotation matrices which are multiplied together to form another rotation.

Confusion upon confusion:
Based upon the match between the 2-dimensional complex space and the 3-dimensional real space, (19.11), We take it that the generators form a 3-dimensional linear space. Quaternion space has no 3-dimensional sub-space, and so

this bit is clearly wrong. None-the-less, the 3-dimensional linear space has an inner product of the form:

$$J^2 = j_1^2 + j_2^2 + j_3^2 \qquad (19.14)$$

Wherein we have denoted the generators by j_i. We call this inner product a Casimir invariant.

On to $SU(3)$:

We have now muddled ourselves into believing in a 3-dimensional linear space of hermitian 2×2 matrices. What about the 3×3 hermitian matrices; there are eight linearly independent 3×3 hermitian matrices. Surely, this must be $SU(3)$. It is in the conventional view of Lie algebras, but it does not correspond to any division algebra. In fact, there are only two unitary division algebras, the Euclidean complex numbers and the quaternions (two types). There can be no unitary Lie algebra corresponding to a division algebra other than $U(1)$ and $SU(2)$.

Summary:

$SU(2)$ is really the quaternion rotation matrix, and a two component complex vector is really a quaternion. All the rest is smoke and mirrors.

Chapter 20

Concluding Remarks

We have seen how easily and sweetly the Lie groups $U(1)$ and $SU(2)$ fall out of the finite groups C_2 and $C_2 \times C_2$. We have seen much other stuff like our 4-dimensional space-time and the Lorentz group $SO(3,1)$ with its sub-groups, the special orthogonal groups, fall out of the same finite groups. Elsewhere, it has been shown that general relativity, the expanding universe with apparent dark energy, and classical electromagnetism fall out of the $C_2 \times C_2$ group. We have seen special relativity within the C_2 group as rotation in 2-dimensional space-time. We have seen that the Weyl spinor, presented as two complex numbers, is really a quaternion in obscure notation. The Weyl spinor represents an electron; we have seen the intrinsic spin properties of the electron (only up or down never any other direction) derived directly from the $C_2 \times C_2$ group as a continuous rotation group, Lie group, in quaternion space. We have even seen left-chiral and right-chiral electrons emerge from the $C_2 \times C_2$ group.

There is much physics, perhaps most of the universe within the finite groups C_2 and $C_2 \times C_2$, but there is no $SU(3)$ Lie group. Indeed, as a Lie algebra, $SU(3)$ is messy. No self-respecting finite group would have anything to do with it.

Yet, $SU(3)$ is a central part of the standard model of quantum field theory.

Your author's view is that nature would not produce such a sweet $SU(2)$ and alongside it such an ugly $SU(3)$. There must be something to replace $SU(3)$. Your author does not have that replacement, and so he must accept $SU(3)$ for now.

We have approached Lie theory from the Lie group side rather than from the Lie algebra side. Your author hopes this approach has enlightened the reader and that the reader has found this approach both gentler and easier than that presented in conventional texts.

This book is primarily concerned with Lie groups. There is more to Lie algebra and Lie groups than has been presented within this short book; for example, we have not considered Dynkin diagrams or Cartan sub-algebras. Nor have we mentioned weights and roots. These concepts are not really anything to do with Lie groups but are within Lie algebras.

If the reader wishes to study Lie algebra further, your author hopes this book will have made her task easier and more intuitively clear.

Dennis Morris

Port Mulgrave

March 2016.

Other Books by the Same Author

The Naked Spinor – a Rewrite of Clifford Algebra

Spinors exist in Clifford algebras. In this book, we explore the nature of spinors. This book is an excellent introduction to Clifford algebra.

Complex Numbers The Higher Dimensional Forms – Spinor Algebra

In this book, we explore the higher dimensional forms of complex numbers. These higher dimensional forms are connected very closely to spinors.

Upon General Relativity

In this book, we see how 4-dimensional space-time, gravity, and electromagnetism emerge from the spinor algebras. This is an excellent and easy-paced introduction to general relativity.

From Where Comes the Universe

This is a guide for the lay-person to the physics of empty space.

Empty Space is Amazing Stuff – The Special Theory of Relativity

This book deduces the theory of special relativity from the finite groups. It gives a unique insight into the nature of the 2-dimensional space-time of special relativity.

The Nuts and Bolts of Quantum Mechanics

This is a gentle introduction to quantum mechanics for undergraduates.

Quaternions

This book pulls together the often separate properties of the quaternions. Non-commutative differentiation is covered as is non-commutative rotation and non-commutative inner products along with the quaternion trigonometric functions.

The Uniqueness of our Space-time

This book reports the finding that the only two geometric spaces within the finite groups are the two spaces that together form our universe. This is a startling finding. The nature of geometric space is explained alongside the nature of division algebra space, spinor space. This book is a catalogue of the higher dimensional complex numbers up to dimension fifteen.

Lie Groups and Lie Algebras

This book presents Lie theory from a diametrically different perspective to the usual presentation. This makes the subject much more intuitively obvious and easier to learn. Included is perhaps the clearest and simplest presentation of the true nature of the Lie group $SU(2)$ ever presented.

The Physics of Empty Space

This book presents a comprehensive understanding of empty space. The presence of 2-dimensional rotations in our 4-dimensional space-time is explained. Also included is a very gentle introduction to non-commutative differentiation. Classical electromagetism is deduced from the quaternions.

The Electron

This book presents the quantum field theory view of the electron and the neutrino. This view is radically different from the classical view of the electron presented in most schools and colleges. This book gives a very clear exposition of the Dirac equation including the quaternion rewrite of the Dirac equation. This is an excellent introduction to particle physics for students prior to university, during university and after university courses in physics.

The Quaternion Dirac Equation

This small book (only 40 pages) presents the quaternion form of the Dirac equation. The neutrino mass problem is solved and we gain an explanation of why neutrinos are left-chiral. Much of the material in this book is drawn from 'The Electron'; this material is presented concisely and inexpensively for students already familiar with QFT.

An Essay on the Nature of Space-time

This small and inexpensive volume presents a view of the nature of empty space without the detailed mathematics. The expanding universe and dark energy is discussed.

Elementary Calculus from an Advanced Standpoint

This book rewrites the calculus of the complex numbers in a way that covers all division algebras and makes all continuous complex functions differentiable and integrable. Non-commutative differentiation is covered. Gauge covariant differentiation is covered as is the covariant derivative of general relativity.

Even Mathematicians and Physicists make Mistakes

This book points out what seems to be several important errors of modern physics and modern mathematics. Errors like the misunderstanding of rotation, the failure to teach the higher dimensional complex numbers in most universities, and the mathematical inconsistency of the Dirac equation and some casual errors are discussed. These errors are set in their historical circumstances and there is discussion about why they happened and the consequences of their happening. There is also an interesting chapter on the nature of mathematical proof within our society, and several famous proofs are discussed (without the details).

Finite Groups – A Simple Introduction

This book introduces the reader to finite group theory. Many introductory books on finite groups bury the reader in geometrical examples or in other types of groups and lose the central nature of a finite group. This book sticks firmly with the permutation nature of finite groups and elucidates that nature by the extensive use of permutation matrices. Permutation matrices simplify the subject considerably. This book is probably unique in its use of permutation matrices and therefore unique in its simplicity.

The Left-handed Spinor

This book covers the left-handed parts of mathematics which we call the chiral algebras. These algebras have CP invariance, violation of parity, and many other aspects which makes them relevant to theoretical physics. It is quite a revelation to discover that mathematics is left-handed.

Other Books by the Same Author

Non-commutative Differentiation and the Commutator

(The Search for the Fermion Content of the Universe)

This book develops the theory of non-commutative differentiation from the fundamentals of algebra. We see what an algebraic operation (addition, multiplication) really is, and we discover that the commutator is a third fundamental algebraic operation within some division algebras. This leads to the first part of the derivation of the fermion content of the universe.

Index

Index

9 781530 187607